Practical Earth Science Exercises

Third Edition

Jonathan A. Nourse
Ernest W. Roumelis
Jeffrey S. Marshall
David R. Berry

California State Polytechnic University
Department of Geological Sciences
Pomona, California

Kendall Hunt
publishing company

We dedicate this book to our friend and coauthor, David Berry, who passed away April 11, 2017 after a long illness. Dr. Berry was a dynamic Professor in the Cal Poly Pomona Geology Department from 1982 to 2014. He was also one of the original authors on the first edition of this book. We have many fond memories of Dave, and will greatly miss his overall optimism, kind spirit, and his enthusiasm for Geology.
Jon Nourse, Jeff Marshall, and Ernie Roumelis

Cover Images: Landslide closes connector from westbound Interstate 10 to northbound 57 Freeway near Cal Poly Pomona campus on February 18, 2010. Photograph shows view to the east; cross section illustrates geologic and groundwater conditions before and after the slide as viewed to the north. Blue lines depict groundwater infiltration to base of conglomerate unit where slide surface developed. This cover image integrates several themes addressed in our "Practical Earth Science Exercises" laboratory manual. Photo and drawing by Jon Nourse, 2010.

Cover image © Photo and drawing by Jon Nourse, 2010.

publishing company

www.kendallhunt.com
Send all inquiries to:
4050 Westmark Drive
Dubuque, IA 52004-1840

Copyright © 1996, 2003, 2017 by Kendall Hunt Publishing Company

ISBN 978-1-5249-3551-1

Published in the United States of America

Contents

Preface

We are pleased to present the 3rd edition of _Practical Earth Science Exercises_, a laboratory manual originally conceived by Drs. Lisa Rossbacher, David Jessey, and David Berry (1986) during their early teaching years at Cal Poly Pomona. Exercises #2-6 and #9-10 have been modified from their original works, following a thorough testing by thousands of Cal Poly Pomona students. Jon Nourse contributed Exercises #1, #11-16, #18-19, and #23 to the 2nd edition; these are extensively revised here. David Berry has revised Exercises #7-8 on Petroleum Geology and Fossil Preservation from the first edition. Jeff Marshall contributed the chapters on Rivers/Floods Hazards (#17) and Neotectonics (#20-21). Exercise #14 was modified from materials used in Marshall's Environmental Geology course at Franklin and Marshall College. Finally, new Exercises #22, #24, and #25 were written by Ernie Roumeli to address the types of geotechnical issues encountered in his Engineering Geology courses. We have reorganized the chapter sequence so that fundamental concepts and definitions are thoroughly introduced before exercises involving practical applications, natural hazards, geotechnical analysis.

Exercises in this volume will acquaint students with practical aspects of physical geology, natural hazards, and engineering applications of geology. The scope encompasses material such as world population distribution, classification of solid Earth materials, topographic map interpretation, construction of geologic maps and cross sections, recognition and assessment of volcanic earthquake, flood, and landslide hazards, measurement of earthquake location, magnitude, and intensity, geologic occurrence of petroleum and groundwater resources, groundwater flow dynamics, and adverse effects of groundwater withdrawal. Topics are chosen to compliment lecture materials covered concurrently in introductory level Earth Science, Natural Disasters, and Engineering Geology courses. Instructors may select various exercises in different sequences to meet the needs of their lecture itinerary. Each exercise is designed to provide the student with "hands on" experience in geological techniques. Some exercises may take less than 50 minutes to complete; others may take three hours. Students will find it most useful to read the introductory material for the appropriate exercise before class meetings. Important terms and definitions ("jargon") are highlighted within the text by _underlined italics_. Students are also encouraged to surf the various excellent web pages cited in the exercises pertaining to natural disasters.

In most cases a short lecture by the instructor will be sufficient to clarify the material contained in this book. For convenience, work to be submitted by students is easily detached from the binder. Mineral and rock specimens, basic identification tools, and maps needed for Exercises #2-6, #8, and #11, are generally available in most geological science laboratories. Many instructors also substitute or add samples from their own prize collections, so take especially good notes on these! Most exercises may be completed by writing directly in this laboratory manual. Additional materials each student should bring to lab are listed at the beginning of each exercise. In general, students will find it most convenient to keep the following items handy:

pencil	eraser	note paper	colored pencils	magnifying glass
ruler	protractor	calculator	graph paper	tracing paper
stapler	compass (for drawing circles)			black, blue, green, and red ink pens

We hope that students find these exercises to provide a practical and educational experience. Enjoy!

Jon Nourse
Ernie Roumelis
Jeff Marshall
Dave Berry

Acknowledgments

Several exercises in this laboratory manual are modified from problems originally conceived by Lisa A. Rossbacher, and David R. Jessey (see above). We gratefully acknowledge their contributions. Some illustrations are reproduced from different environmental geology/geologic hazards textbooks; others are modified from published scientific journals. In such cases, citations given in the figure captions are linked to the bibliography at the end of the manual. Several web sites that provided very helpful information for exercises pertaining to natural disasters are cited at relevant places in the text. We are grateful to Rosalie Giroux, Bernice Gilbert, and Sharon Cruise for their patience in scanning, typing and formatting earlier versions of this book. The Geological Sciences Department at Cal Poly Pomona provided the materials and computer facilities needed to produce a camera-ready manuscript.

Name: _____

Class Number: _____

Plate Boundaries and Global Population Distribution

Objectives: To acquaint the student with locations of active plate boundaries and the geographic distribution of global population centers.

Materials: ruler
colored pencils: red, green, dark blue, light blue, brown
black ink pen

Introduction

Current estimates (2017; **Figure 1.1**) place Earth's human population at 7.5 billion inhabitants (up from ~6.1 billion in 2000; ~4.4 billion in 1980). This exercise utilizes locations of Earth's thirty most populous cities as a measure of the general distribution of global population. In the course of plotting these cities on a geologic map, students will become familiar with general world geography and might note a clustering of cities in particular geologic settings. Spatial associations may be drawn between certain population centers and important structural/tectonic features known as _plate boundaries_.

Figure 1.1. List of the World's 100 Most Populous Cities*, Ranked by Metropolitan Area

Rank	City	Nation	City Proper	Metropolitan Area	Urban Area	Rank	City	Nation	City Proper	Metropolitan Area	Urban Area
1	Chongqing	China	8,189,800	52,100,100	36,700,000	51	Barcelona	Spain	1,604,555	5,375,774	4,740,000
2	Guangzhou	China	13,080,500	44,259,000	20,800,654	52	Visakhapatnam	India	2,035,922	5,340,000	
3	Tokyo	Japan	13,513,734	37,843,000	36,923,000	53	Xiamen	China	3,531,347	5,114,758	1,861,289
4	Shanghai	China	24,256,800	34,750,000	23,416,000	54	Guayaquil	Ecuador	3,600,000	5,000,000	
5	Jakarta	Indonesia	10,075,310	30,539,000	30,075,310	55	Quito	Ecuador	2,671,191	4,700,000	
6	Karachi	Pakistan	23,500,000	25,400,000	25,400,000	56	Ankara	Turkey	5,271,000	4,585,000	4,919,000
7	Delhi	India	16,787,941	24,998,000	21,753,486	57	Phoenix	United States	1,563,025	4,574,531	3,629,114
8	Beijing	China	21,516,000	24,900,000	21,009,000	58	Addis Ababa	Ethiopia	3,103,673	4,567,857	3,384,569
9	São Paulo	Brazil	12,038,175	21,090,791	36,842,120	59	Guadalajara	Mexico	1,495,189	4,424,252	
10	New York City	United States	8,550,405	20,182,305	23,723,696	60	Rome	Italy	2,874,038	4,362,282	
11	Mexico City	Mexico	8,974,724	20,063,000	21,178,959	61	Kochi	India	2,232,456	4,221,140	
12	Osaka	Japan	2,691,742	19,341,976	17,444,000	62	Montreal	Canada	1,649,519	4,127,100	3,407,963
13	Cairo	Egypt	10,230,350	18,290,000	22,439,541	63	Medan	Indonesia	2,097,610	4,103,696	
14	Mumbai	India	12,478,447	17,712,000	20,748,395	64	Chittagong	Bangladesh	2,581,643	4,009,423	
15	Moscow	Russia	12,197,596	16,170,000	16,800,000	65	Salvador	Brazil	2,902,927	3,919,864	
16	Dhaka	Bangladesh	16,970,105	15,669,000	18,305,671	66	Brasília	Brazil	2,556,149	3,919,864	
17	Los Angeles	United States	3,884,307	15,058,000	13,262,220	67	Faisalabad	Pakistan	6,480,765	3,675,000	
18	Bangkok	Thailand	8,280,925	14,998,000	8,305,218	68	Cologne	Germany	1,057,327	3,573,500	
19	Kolkata	India	4,486,679	14,667,000	14,617,882	69	Milan	Italy	1,359,905	3,206,465	
20	Buenos Aires	Argentina	3,054,300	14,122,000	13,074,000	70	İzmir	Turkey	4,168,000	3,019,000	3,575,000
21	London	United Kingdom	8,673,713	13,879,757	9,787,426	71	Budapest	Hungary	1,759,407	2,927,944	3,303,786
22	Tehran	Iran	8,154,051	13,532,000	14,595,904	72	Munich	Germany	1,450,381	2,606,021	
23	Istanbul	Turkey	14,025,000	13,520,000	14,657,000	73	Vienna	Austria	1,863,881	2,600,000	
24	Kinshasa	Congo	10,130,000	13,265,000		74	Sapporo	Japan	1,918,096	2,584,880	
25	Lagos	Nigeria	16,060,303	13,123,000	21,000,000	75	Davao City	Philippines	1,632,991	2,516,216	
26	Manila	Philippines	1,780,148	12,877,253	24,123,000	76	Nagpur	India	2,405,665	2,497,870	
27	Rio de Janeiro	Brazil	6,429,923	12,727,000	13,973,505	77	Bucharest	Romania	1,883,425	2,272,163	1,931,000
28	Seoul	South Korea	9,995,784	12,700,000	25,520,000	78	Belgrade	Serbia	1,166,763	1,659,440	1,344,844
29	Shantou	China	5,391,028	11,535,677		79	Auckland	New Zealand	1,495,000	1,614,300	
30	Tianjin	China	6,859,779	10,920,000		80	Tbilisi	Georgia	1,118,035	1,485,293	
31	Lima	Peru	8,852,000	10,750,000	9,886,647	81	Calgary	Canada	1,235,171	1,214,839	1,095,404
32	Shenzhen	China	10,778,900	10,630,000	12,084,000	82	Tbilisi	Georgia	1,118,035	1,485,293	
33	Paris	France	2,229,621	10,601,122		83	Suzhou	China	10,650,501		
34	Chengdu	China	4,741,929	10,376,000		84	Dongguan	China	8,220,207		
35	Lahore	Pakistan	10,052,000	10,355,000		85	Nanjing	China	8,187,828		
36	Bengaluru	India	8,425,970	9,807,000	8,728,906	86	Shenyang	China	8,106,171		
37	Bogotá	Colombia	7,878,783	9,800,000	9,520,000	87	Hyderabad	India	7,859,250		8,955,450
38	Chicago	United States	2,695,598	9,156,000	9,554,598	88	Ho Chi Minh City	Vietnam	7,681,700		
39	Nagoya	Japan	2,296,014	9,107,414	10,177,000	89	Baghdad	Iraq	7,180,889		
40	Busan	South Korea	3,510,833	8,202,239		90	Fuzhou	China	7,115,369		4,468,076
41	Hong Kong	Hong Kong	7,298,600	7,331,699		91	Chennai	India	7,088,000		9,373,521
42	Surabaya	Indonesia	2,765,487	7,302,283		92	Changsha	China	7,044,118		
43	Bandung	Indonesia	2,575,478	6,965,655		93	Wuhan	China	6,886,253		
44	Houston	United States	2,489,558	6,490,180	4,944,332	94	Hanoi	Vietnam	6,844,100		
45	Toronto	Canada	2,731,571	6,417,516	6,129,900	95	Foshan	China	6,151,622		
46	Quanzhou	China	8,128,533	6,107,475	1,435,185	96	Zunyi	China	6,127,009		
47	Philadelphia	United States	1,567,442	6,051,170	7,146,706	97	Santiago	Chile	5,743,719		6,683,852
48	Saint Petersburg	Russia	5,191,690	5,900,000	6,200,000	98	Riyadh	Saudi Arabia	5,676,621		
49	Berlin	Germany	3,517,424	5,871,022		99	Ahmedabad	India	5,570,585		6,352,254
50	Fukuoka	Japan	1,483,052	5,590,378		100	Singapore	Singapore	5,535,000		

Source: Wikipedia, 2017; see https://en.wikipedia.org/wiki/List_of_largest_cities#Urban_area for details of data sources

Plate boundaries are discontinuities that define the margins of the lithosphere plates. Plate movements are responsible for many of the geologic processes that shape Earth's outer surface. Your instructor has undoubtedly described some details of plate movements and their consequences during lecture. Many of Earth's internally driven catastrophic processes such as volcanic eruptions, earthquakes, tsunami, and landslides tend to occur in active plate boundary regions. The following exercise of coloring a *plate-tectonic map* of the world (**Figure 1.2**) and locating major cities will help you appreciate spatial relationships between global population and plate boundary locations. To statistically analyze these relationships you will create a *histogram* to show the number of cities located at specific distances from plate boundaries.

Procedure / Problems

1. Color your map of Earth and its plates (**Figure 1.2**), using the following color scheme:
> *land masses*: **light brown** (shade these areas lightly)
> *oceans and seas*: **light blue** (shade these areas lightly)
> *convergent plate boundaries*: **dark blue** lines
> *divergent plate boundaries*: **dark red** lines
> *transform plate boundaries*: **dark green** lines

2. Carefully locate the 30 most populated cities of the world (**Figure 1.1**), shown as **distinct black squares** on your map (**Figure 1.2**). Precise locations may be obtained from internet search of maps for the various countries. Label each city with a one- or two-letter abbreviation (e.g., **LA** for Los Angeles or **NY** for New York). Use black ink.

3. Use a scrap of paper to fashion a *simple scale* that corresponds to the markings of the bar scale on the map (your instructor will demonstrate this). Use your scale to determine the distance from each city to the closest plate boundary in **kilometers**. Keep track of each distance for Problem 4 below.

4. Use the template of **Figure 1.3** to make a **histogram** of *number of cities* vs. *distance to nearest plate boundary*. Subdivide your horizontal axis into five groupings as follows: 0-500 km, 0-1000 km, 1001-2000 km, 2001-3000 km, and greater than 3000 km. Each city will plot as a box with height of 1 unit on the vertical axis. Your completed chart will be a bar graph, with five bars extending upward from the horizontal axis. The height of each bar will depend on the number of cities contained in the specified distance range. Each bar should be composed of stacks of individual boxes. *For example*, Los Angeles is about 80 km from a transform plate boundary (the San Andreas fault). Tokyo, Japan is about 250 km from a convergent plate boundary. Seoul, South Korea is about 1800 km from a transform plate boundary. Each of these cities plots as a separate box as shown on Figure 1.3. As you draw your boxes, label each with the name of the corresponding city.

5. Explain your histogram and some general geographic patterns by answering the following questions in the spaces below:
> **a.** What does this chart say about the general distribution of Earth's population with respect to proximity to active plate boundary regions?

> **b.** Consider those cities that are within 500 km of an active plate boundary. How many are close to a convergent plate boundary? Name them. How many are close to a transform plate boundary? Name these. How many are close to a divergent plate boundary? Name them.

> **c.** What percentage of your 30 cities are located on coastlines? Considering what you know about world history, how would you explain this relatively large number?

> **d.** How many of your coastal cities are located within 500 km of a plate boundary? How many are within 1000 km?

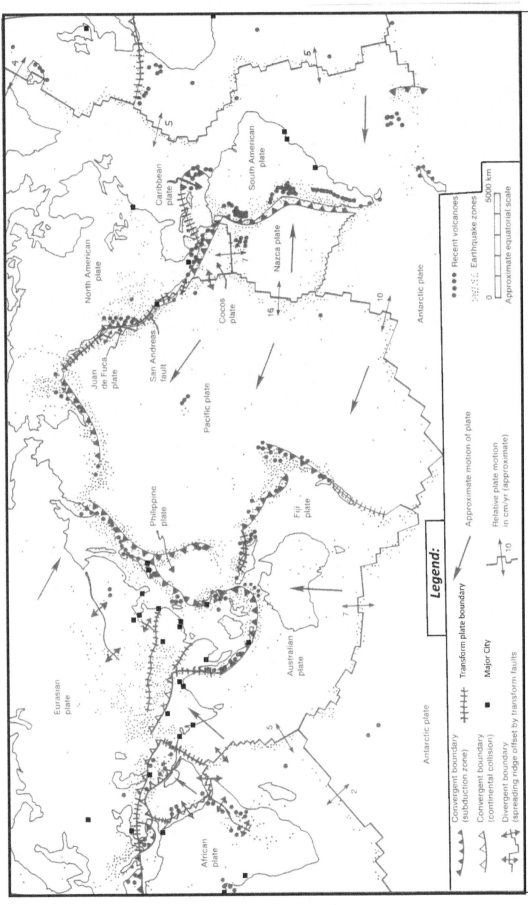

Figure 1.2. Map showing Earth's lithospheric plates, various plate boundaries, and relative plate motions (modified from Figure 3.2 in Keller, 1992)

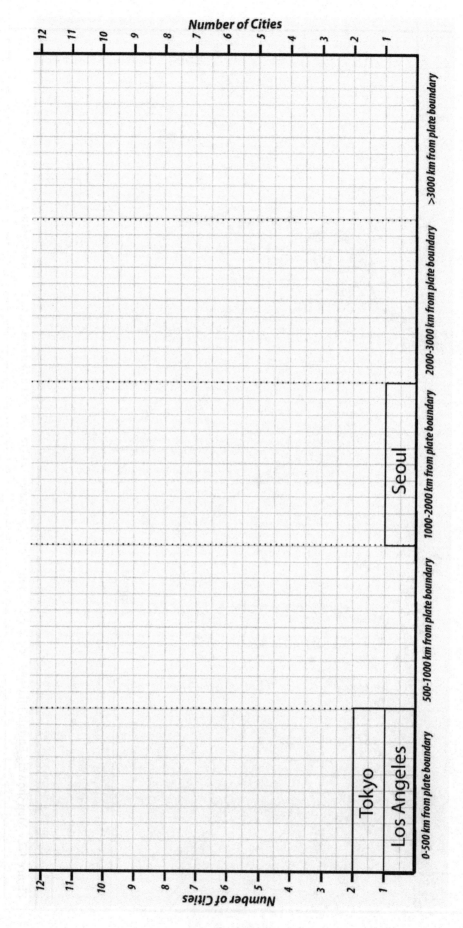

Figure 1.3. Template for creating Histogram related to Problem 4.

4

Mineral Properties and Identification

Objectives: (a) To introduce the properties and terminology used to describe the characteristics of minerals.
               (b) To easily identify and classify common rock forming minerals.

Materials:
Pencil
Textbook
Hardness testing kit provided by instructor
Other tools made available in laboratory; e.g., porcelain streak plate, glass scratch plates, magnet, magnifying lens
Mineral Identification Sheet provided at end of this chapter
Internet access to mineral web sites: http://geology.com/minerals/ ; http://www.webmineral.com/ ; http://www.minerals.net/

Introduction

Geologists generally consider minerals to be the basic building blocks of all inorganic Earth materials. They are the natural crystalline solids that make up nearly all rocks within the Earth's crust. Each mineral type consists of a different chemical compound with a unique arrangement of atoms that gives the mineral specific characteristics by which it can be identified. All minerals have the following characteristics:

1. naturally occurring solids (not man-made)
2. inorganic (not composed of living matter)
3. crystalline (unique internal arrangement of atoms)
4. of fixed chemical composition (can be expressed by a chemical formula)

During subsequent laboratories, you will be asked to identify some common rock-forming minerals. Over 2000 different minerals occur in nature, but relatively few of these are common. How do you go about the process of identification? There are a number of techniques for mineral identification: (1) x-ray diffraction, (2) chemical analysis, (3) crystallography, and (4) direct observation of physical properties. The first three methods are time consuming and require expensive equipment or special training. The fourth method, although not definitive in some instances, works well for the more common minerals.

What are physical properties? Simply, they are any characteristic that a mineral possesses that can be readily observed. The following section will introduce you to some of the most common physical properties which can be used in subsequent exercises for mineral identification. ***Please refer to the web sites listed above for additional information***.

Description of Physical Properties

Mineral Color
Color may be used to identify some minerals, but it is often a confusing and misleading property. Examine the collection of mineral specimens you are given. How many different colors of each can you distinguish? You should see why color cannot be considered diagnostic for these as well as many other minerals.

Mineral Streak
Streak, the color of a powdered mineral, is considerably more useful for identification purposes than the color of the solid mineral. Streak is found by rubbing the mineral on a piece of unglazed porcelain tile. Not all minerals will leave a streak.

Mineral Luster
Luster refers to the appearance of a mineral in reflected light. We will concentrate mostly on the difference between _metallic_ and _nonmetallic_ lusters. Any metallic minerals should have a metallic luster; you can see this luster by holding a fresh penny or a piece of metal jewelry to the light and seeing the reflection of light off the surface. Be careful not to confuse the glassy, nonmetallic luster of some minerals with those that have metallic luster. If you can see through the mineral or a thin piece broken from the mineral, it does not have a metallic luster. Words used to describe nonmetallic luster include _vitreous_ (glassy), _resinous, pearly, greasy, silky_, and _dull_ (earthy).

Mineral Hardness

Hardness refers to the resistance of a mineral to scratching or abrasion. The **Mohs Hardness Scale**, given below, lists some common minerals in order of increasing hardness. You are expected to be familiar with it. A number of hardness sets are available in the lab for you to use. The standard reference minerals for Mohs Hardness Scale are:

1. Talc	**6.** Orthoclase
2. Gypsum	**7.** Quartz
3. Calcite	**8.** Topaz
4. Fluorite	**9.** Corundum
5. Apatite	**10.** Diamond

Typical materials to help you determine hardness are a piece of glass (H=5.5), a steel nail (H~5.5), unglazed tile (H~5), a copper penny (H=3), and your fingernail (H~2.5).

Mineral Cleavage

Many minerals have an internal array of atoms that are bonded more strongly in some directions than others. These minerals often break along the weaker planes, resulting in surfaces that are generally smooth. These surfaces will usually reflect light as the mineral is rocked back and forth under a bright lamp. See your text for a further discussion of cleavage. Minerals may have more than one "direction" of cleavage. Be careful not to confuse cleavage surfaces (planes), which are breakage phenomena, with flat crystal faces that form during the crystal's growth. Many cleavage planes parallel to crystal surfaces, but not all are. Quartz and garnet can be tricky, because they form beautiful crystal faces during growth, but they break along rough, irregular surfaces (*conchoidal fracture*).

Minerals that do not possess cleavage are said to have *fracture*. This means they have no tendency to split (cleave) along anyone preferred plane.

Procedure

You are now prepared to identify the unknown mineral samples chosen by your instructor. You may be expected to identify many of these minerals during a quiz in the upcoming weeks, so pay close attention to what you are doing and don't let your lab partner do all the work. Begin with **Table 2.1a** below, then proceed to the appropriate **Mineral Identification Keys 2A-2C** provided on pages 7-8. As you identify the minerals, fill out the ***Mineral Identification Worksheet*** on page 10. You will need to refer to **Table 2.1b** on page 9 to find the chemical composition of the minerals. Information contained in the following keys and charts will be useful. The basic procedure entails a process of elimination.

Table 2.1a: Flow Chart to the Mineral Identification Keys

Nonmetallic luster	if light go to:	KEY 2A, and test hardness
	if dark go to:	KEY 2B, and test hardness
Metallic luster	if light or dark go to:	KEY 2C, and test streak

---

Exercise #3

Name: _____

Class Number:_____

Identifying Igneous Rocks

Objective: To identify and classify common igneous rocks containing known rock-forming minerals.

Materials: Exercise #2
textbook
tools made available in laboratory, glass scratch plates, hand lenses, binocular microscopes
pencil
Igneous Rock Identification Worksheet provided at end of this chapter
Internet access to igneous rock web sites: http://geology.com/rocks/igneous-rocks.shtml
https://en.wikipedia.org/wiki/Igneous_rock

Introduction

Igneous rocks form by cooling of a molten silicate liquid called a magma. The basis for classification of igneous rocks is usually _texture_ and _mineralogy_. In the latter case, however, mineral grains can often be too fine grained to be identified, or, in fact, can be absent altogether.

Texture can be defined as the size and shape of the constituent mineral grains. There are only four basic igneous rock textures. _Coarse-grained_ or "phaneritic" texture is the term used when an igneous rock contains large, intergrown crystals, all of which can be identified with the eye or microscope. Phaneritic rocks form in an environment where slow cooling allows for maximum crystal growth, generally at great depth beneath the Earth's surface. _Fine-grained_ or "aphanitic" rocks also consist of interlocking crystals, but in general the crystals are too small to be identified with anything but the most powerful microscope. The fine grain size of aphanitic rocks indicates an environment where the rocks cooled too quickly to allow for large crystal growth. Typically, these rocks form at or near the Earth's surface. Because the individual crystals in aphanitic rocks are often too small to be seen with the eye, we are often forced to rely on color to make an identification. Keep in mind, however, that color can be as unreliable a property in rocks as it is in minerals. Fortunately, under special circumstances to be discussed below, we can often identify a few large crystals in what is basically a fine-grained rock. These few key mineral grains allow us to name the aphanitic rock even though we can't see or identify most of the crystals that make up the rock.

An additional textural term of importance for crystalline igneous rocks is _porphyritic_. Porphyritic textured rocks consist of two conspicuously different grain sizes. The difference in grain size indicates two different environments of formation. An example might be a magma that cooled very slowly at depth for hundreds of thousands of years and then suddenly escaped to the Earth's surface to complete its solidification. Grains that formed at depth would be coarse and those cooling at the Earth's surface would be fine. Hence, the texture of the rock is porphyritic.

The significance of a porphyritic texture lies in its usefulness when identifying aphanitic rocks. Because aphanitic mineral grains are too small to be seen, we often look carefully for even a few larger crystals, termed phenocrysts, in the aphanitic groundmass. If we can identify these phenocrysts, it often allows us to provide a far more reliable name for the rock than we could if we based the name on color alone. Hence, porphyritic texture is particularly important in the aphanitic rocks because it usually provides the only key we have to confidently name the rock. Keep in mind, however, that porphyritic texture is not confined to aphanitic rocks but can also occur in phaneritic rocks even though it is far less common in this group.

The remaining igneous rock textures, _glassy_ and _fragmental_, do not apply to crystalline rocks, but are used for rocks that cooled too rapidly to allow crystallization to begin. Glassy textured rocks are, as the name implies, true glasses. They reflect light and have a shiny appearance, and they can also be porous and light in weight. Fragmental textured rocks are also glasses but, in reality, consist of a mosaic of tiny fragments. They do not appear shiny like a glass and, hence, are more difficult to recognize.

Procedure

How do you go about identifying an igneous rock? The first step is to examine your rock sample and determine its texture. Can you see each mineral grain, only a few, or none at all? To be phaneritic in texture, you must be able to point to each individual crystal, even if you don't know what they are! If you see only a few recognizable mineral grains in a dull groundmass, you are probably dealing with an aphanitic rock. Glassy textured rocks are either quite shiny or very light-weight. Fragmental rocks consist of angular fragments, often loosely bound. Let's say you have decided your unknown rock is phaneritic in texture. What next?

The next step involves identification of the visual mineral grains. Refer to **Figure 3.1** and **Table 3.1** to assist you with this identification, along with **Keys 3A – 3E**. Note that there are only eight important minerals in igneous rocks. Let's say that we can identify quartz, orthoclase, and plagioclase in our sample. Now all we need to do is go to one of the two identification charts and read off the name, in this case Granite.

Using the tables and keys attached to this exercise, please fill out the **Igneous Rock Identification Worksheet** on page 15. Please note that rocks with a glassy or fragmental texture cooled too quickly to develop individual minerals; "mineral composition" should therefore be described as "same as the silicate magma." In the "origin" column, specify "intrusive" or "extrusive" and provide a plate tectonic setting, as described by your instructor.

	Cooling Rate	Texture	Light Color	Intermediate Color (gray)	Dark Color	Olive Green or Black
Surface of Earth	Very fast cooling	Glassy	Obsidian			
		Glassy with Vesicles	Pumice		Scoria	
		Fragmental	Rhyolite tuff		Basalt tuff	
	Fast Cooling	Fine Grained (aphanitic)	Rhyolite	Andesite	Basalt	Komatiite
Deep	Slow cooling	Coarse Grained (phaneritic)	Granite	Diorite	Gabbro	Dunite or Peridotite

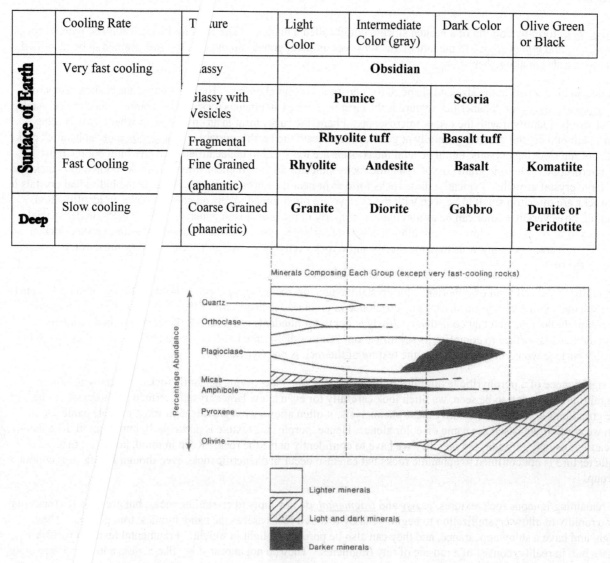

Minerals Composing Each Group (except very fast-cooling rocks)

Figure 3.1. Texture, rate of cooling, and mineral composition of igneous rocks. These two charts correspond; for example, a light-colored coarse-grained rock will be identified as a granite. Reading down the column to the lower chart the mineral composition of granite can be found.

Exercise #4

Name:_____

Class Number:_____

Identifying Sedimentary Rocks

Objectives: To identify and understand the origin of sedimentary rocks through observations about their texture and composition.

Materials: Exercise #2
textbook
tools made available in laboratory: glass scratch plates, hand lenses, binocular microscopes, dilute HCl
pencil
Sedimentary Rock Identification Worksheet provided at end of this chapter
Internet access to igneous rock web sites: http://geology.com/rocks/sedimentary-rocks.shtml
https://en.wikipedia.org/wiki/Sedimentary_rock

Introduction

Sedimentary rocks cover nearly 75% of the Earth's land surface. They are composed of the accumulated fragments of previously formed rocks, minerals, and organic materials. These deposits are subsequently consolidated by cementation and compaction. Sedimentary rocks can also form from other Earth materials that have been dissolved and which are later precipitated or incorporated into organisms. All these rocks are characterized by a layered structure that results from variations in the constituent minerals or grains as they were deposited. Fossils, the remains of previously living organisms, are found almost exclusively in sedimentary rocks.

Sedimentary rocks are classified on the basis of their grain sizes and _texture,_ in addition to their _composition._ The two major textures for sedimentary rocks are _clastic_ and _chemical._ Clastic rocks form as a result of the accumulation of fragments of pre-existing rocks (igneous, metamorphic, or other sedimentary rocks). These older rocks were exposed to weathering and erosion, and then the individual sediments were transported to the site of deposition. An example of this type of sedimentary rock texture is a sandstone. Chemical sedimentary rocks form by either purely chemical precipitation or the accumulation of organic materials. Rock salt (composed of the mineral halite) and some limestone (composed of the mineral calcite) are examples of pure chemical rocks. _Coal_ and _chert_ are examples of biochemical sedimentary rocks.

Sedimentary Rock Classification

Common Sedimentary Minerals

Many of the minerals that appear in sedimentary rocks are derived from igneous rocks, particularly those minerals that formed late in Bowen's Reaction Series and which are therefore more stable at the Earth's surface. Thus, quartz and feldspars are common constituents of sedimentary rocks. Other sedimentary minerals include calcite, dolomite, halite, and gypsum. These are listed in your key from Exercise #2. In addition to these, sedimentary rocks can also include clay minerals, which are common weathering products of feldspars, and iron oxides, which can color rocks red or yellow. Iron oxides often occur as cement in sandstones, holding the individual sediment particles together.

Clastic Sedimentary Rocks

Clastic rocks are composed of collections of individual grains. These rocks are further classified on the basis of grain size, which is defined by the diameter of the dominant size of particles. In addition to grain size, the following properties are very important in understanding clastic rocks: (1) _composition_ (the minerals or rock fragments present and their abundance), (2) _shape_ (rounded or angular), and (3) _sorting_ (the relative distribution of grain sizes).

The key for identification of clastic rocks begins with the grain size and shape. A coarse-grained clastic sedimentary rock (average particle size = 2 mm or larger) is either a _conglomerate_ or a _breccia_ (rhymes with "betcha"), depending on whether the grains in it are more rounded or more angular. The _roundness_ of the grains depends on the sharpness of their corners and edges. **Figure 4.1** illustrates some of the common terminology for grain roundness.

Medium-grained clastic rocks (particle size = sand) are called *sandstone*. This group of sedimentary rocks is typically subdivided by compaction and sorting. *Sorting* indicates the distribution of grain sizes present. Different degrees of sorting are shown in **Figure 4.2**. The major types of sandstone are:

quartz arenite or *quartzose sandstone* - a "clean," quartz-rich sandstone, characterized by having more than 90% quartz and little or no matrix (very fine-grained silt and clay between the sand particles).

arkose - a feldspar-rich sandstone, with at least 25% feldspar and generally less than 15% matrix; often pink because of the abundant orthoclase feldspar.

greywacke - a "dirty" rock, with at least 15% matrix and generally less than 75% quartz; frequently contains amphibole, biotite, and whole fragments of other rocks.

lithic sandstone - a sandstone containing more rock fragments than feldspar grains.

The smallest grain size (particle size = silt or clay) for a clastic sedimentary rock is found in siltstone and shale. These can sometimes be difficult to identify as a sedimentary rock because the grains are too small to see, but you can recognize them as sedimentary because they break into thin, flat pieces (see the examples in lab). *Siltstone* will usually scratch glass (if you try hard enough) because of all the fine-grained quartz it contains. *Shale* can be easily scratched with glass. Sometimes the name *mudstone* is used for rocks composed of uncertain proportions of silt and clay.

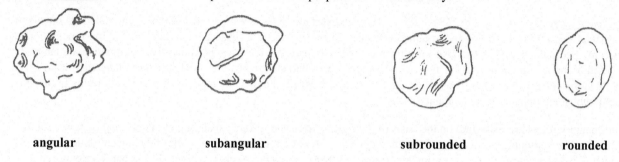

| angular | subangular | subrounded | rounded |

Figure 4.1. Textural terms used by sedimentologists to describe grain roundness.

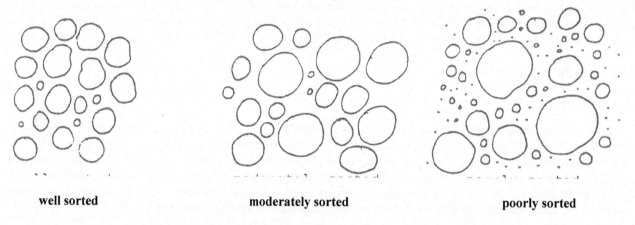

| well sorted | moderately sorted | poorly sorted |

Figure 4.2. Textural terms to describe sorting in sedimentary rocks.

Chemical Sedimentary Rocks

Chemical rocks form by the precipitation of solid material from solution. These rocks are frequently classified on the basis of composition. *Siliceous* rocks contain large amounts of silica (chemical combinations of silicon and oxygen). One example of this is *diatomite*, a soft, white, shale-like rock primarily composed of the siliceous shells (also called "tests") of single-celled algae called diatoms. (Because they are composed of silica compounds, diatoms are very hard. They are often used in "whitening" toothpaste as an abrasive). *Chert* can form through the inorganic precipitation of silica, or from the accumulation of the silica-rich shells of diatoms or radiolarians (microscopic zooplankton). *Evaporite* is a rock formed by the partial or total evaporation of either sea water or ion-rich lake water. Gradual evaporation results in the precipitation of first *calcite* (to form limestone), then gypsum and finally halite (rock salt) and other minerals.

Carbonate rocks (carbonates) are an especially abundant and important class of chemical sedimentary rocks. Basically, these can be divided into *limestones* and *dolostones* (magnesium-rich limestones). Most limestones are organic in origin; they are composed of the shells of microscopic calcareous organisms (*foraminifera*, for example) or the debris from larger organisms (corals, bryozoans, clams, calcareous algae, etc.). Early studies of carbonate rocks were based on grain size, sorting, and roundness but this approach was not very useful. Today, carbonate classification really depends on detailed study using a petrographic microscope to determine the abundance of lime mud (*micrite*). We will rely on more basic observations to identify limestones and dolostones.

Carbonaceous rocks represent accumulations of organic matter, like *peat* or *coal*. Peat is a loosely compacted mass of plant material that formed under reducing conditions. When this peat is subjected to heat and pressures associated with burial, physical and chemical changes convert it into bituminous and then anthracite coal. These rocks are grouped with the chemical Sedimentary rocks, but their texture is classified as "***organic***."

Formation of Sedimentary Rocks

Sedimentary rocks are part of the *rock cycle*, which generally follows these steps:

1. *Exposure* of older rocks (igneous, metamorphic, or sedimentary) at the Earth's surface. These rocks are then weathering by two mechanisms:

Physical weathering includes processes responsible for breaking rocks apart, such as the expansion of freezing water in cracks or the growth of plant roots. Physical weathering increases the surface area available for additional weathering.

Chemical weathering involves the chemical decomposition of pre-existing minerals and rocks. A common example is the weathering of feldspars, which can produce clay minerals and dissolved ions, and may also free small pieces as clastic grains.

2. *Erosion* (transportation) processes then begin to carry the weathered material away from the place where it formed. The agent of erosion may be water, wind, ice (as in glaciers), or gravity (as in landslides and other earth movements). Dissolved materials are carried away in solution to the ocean or are trapped in lakes or desert playas. Clastic particles may be suspended in water by turbulence, or they can be bounced or rolled along the bottom of a stream or across the desert floor.

As these particles bounce and roll, they constantly strike each other, as well as the bottom and walls of whatever they are being transported through or across. These collisions rapidly reduce the particle size until the clasts can be suspended in water. Materials in the process of being transported generally become smaller and more rounded; they usually get sorted into groups of similar-sized clasts at the same time. Minerals that dissolve in water or weather easily are removed progressively, leaving more resistant minerals such as quartz. These changes are expressed as the maturity of the sediment, a general term that indicates the percent of resistant minerals and the degree of sorting and roundness of the grains.

3. *Deposition* describes the way in which the sediments are no longer carried by the transporting agent and are dropped to the Earth's surface. For clastic sediments, the effects of different depositional processes include the sizes of sediment deposited at a particular locality, packing of the grains as they are deposited, and formation of various sedimentary structures.

In general, the larger-sized grains indicate a depositional environment with high energy (rivers and beaches, for example) and smaller grain sizes reflect low-energy depositional environments (ocean floors and quiet lakes, for example). With longer and longer times of exposure to the geologic processes of erosion and transportation, the sediments become (a) smaller, (b) more rounded, and (c) better sorted (closer to being all the same size). As you look at the sedimentary rocks in this exercise, think about what their environment of deposition may have been.

4. After deposition, other changes occur. *Diagenesis* is the process of physical and chemical change following deposition of sediment. The end result of these changes is *lithification*, or the conversion of sediments to "solid rock." These processes may be rapid or extremely slow. Compaction is due to the weight of sediments deposited later, and it may result in dewatering of the sediment, increased cohesiveness of clay particles, and changes in packing. Water moving through the sediments, from a variety of sources, may dissolve some minerals and deposit others. This water can also cause chemical changes in existing minerals, including cementation, which is the precipitation of binding material between and around grains of sediment. Organic material in the sediment may be oxidized and the acidity may change, both of which could cause additional changes in the system.

Finally, if enough heat and pressure are applied, the sedimentary rocks may be changed to *metamorphic rocks* (**see Exercise #5**). Because this is a gradual process, deciding whether a particular rock is sedimentary or metamorphic may be difficult. If the rock is exposed to enough heat and pressure to melt it, the materials are converted to magma and the rock cycle begins again with formation of a new igneous rock when the magma cools.

Depositional Environments

Sedimentary rocks are especially useful to geologists for interpreting *depositional environment*; i.e., the surface geologic setting where the sediments were originally deposited. This is helpful for reconstructing the surface environment that existed millions or billions of years in the past. Commonly the sedimentary rock sequence exposed at a given place (for examples, the Grand Canyon- Zion Canyon- Bryce Canyon succession of northern Arizona and southern Utah) will reveal a complicated history of changing depositional environments. The drawing below **(Figure 4.3)** links the names of common clastic and sedimentary rocks to specific depositional environments. Also shown are conventional patterns used by geologists to depict common sedimentary rocks:

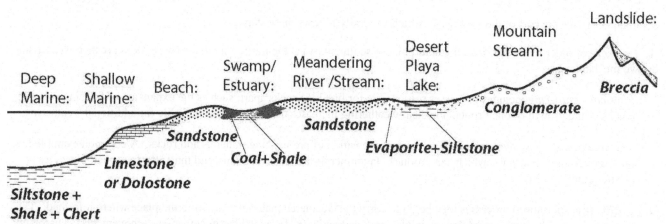

Figure 4.3. Cross section illustrating some common depositional environments and the sedimentary rock types associated with each. Also shown are conventional patterns used by geologists to illustrate these rocks. Drawing by J. Nourse.

Procedure

Identify each of the samples you are given, using **Keys 4A-4C**. Pay particular attention to the texture and the constituent minerals that you can identify in each rock and the texture as you fill out the attached identification tables. On the *Sedimentary Rock Identification Worksheet* (page 22) you should list the minerals that make up the particular rock you are studying under the column labeled "mineral composition." Texture will be either clastic or chemical, based on your observations and use of the chart below. Be sure to specify a *depositional environment* for each sample.

Key 4A. Identification Chart for Sedimentary Rocks

Individual particles visible with the eye or microscope; or feels gritty	Particles scratch glass:	Go to key for Clastic rocks
	Particles do not scratch glass:	Go to key for Chemical rocks
	Reacts with acid (powder if necessary):	Go to key for Chemical rocks
Individual particles not visible with the eye or microscope; or does not feel gritty	Coal black, shiny light weight:	Go to key for Chemical rocks
	White to light gray Does not react with acid:	Go to key for Chemical rocks
	Layered, dense, dull black or green or red:	Go to key for Clastic rocks

Exercise #5

Name: _____

Class Number: _____

Identifying Metamorphic Rocks

Objective: To identify and understand the origin of metamorphic rocks through observations of their texture and composition.

Materials: Exercises #2, #3, and #4
textbook
tools made available in laboratory: glass scratch plates, hand lenses, binocular microscopes, dilute HCl
pencil
Metamorphic Rock Identification Worksheet provided at end of this chapter
Internet access to igneous rock web sites: http://geology.com/rocks/metamorphic-rocks.shtml
https://en.wikipedia.org/wiki/Metamorphic_rock

Introduction

Metamorphic rocks by definition have "changed form" as a result of alteration or recrystallization of pre-existing rocks by agents of metamorphism that include heat, pressure, and chemically active fluids. The pre-existing rocks, also known as _protoliths_, may be igneous, sedimentary, or other metamorphic rocks. Metamorphic rocks can be subdivided into three distinct groups depending on whether their formation involved _regional_, _contact_, or _dynamic metamorphism_.

Regional metamorphic rocks **(Figure 5.1)** form in tectonically active areas in response to the increase in temperature caused by large-scale burial of rock bodies, and also in response to the directed pressure generated during plate movements. As a result of this close association with directed pressure, regionally metamorphosed rocks often possess a foliated texture. _Foliated_ texture is distinguished by a planar arrangement of platy or tabular minerals during recrystallization. The net effect is to give a foliated rock the appearance of layering even though this layering was not generated as the result of sedimentary processes, but rather by directed pressure during metamorphism. One interesting engineering aspect is the tendency for these rocks to split along planes oriented parallel to foliation. This happens because the mineral bonds are weaker (have lower shear strength) along the foliation planes. Thus it is safer to load a foliated metamorphic rock perpendicular rather than parallel to layering.

Contact metamorphic rocks **(Figure 5.1)** form adjacent to shallow cooling plutons (igneous masses deep within the Earth), and in general the important agents of contact metamorphism are heat and chemically active fluids, both supplied by the nearby pluton. Pressure is usually unimportant, and as a result foliation is characteristically absent in contact metamorphic rocks. The net result is an equigranular to distinctly porphyritic (the metamorphic term is porphyroblastic) texture with no indication of any layering. In general, these kinds of textures are classified as _non- foliated_ texture.

Dynamic metamorphism only occurs adjacent to major fault zones. Pressure, mechanical shearing, and grain-size reduction are significant agents of this type of metamorphism. Rocks like _mylonite_ (derived from a word meaning "to mill") and _fault breccia_ form in this environment. Mylonites deserve mention at Cal Poly Pomona because the San Gabriel Mountains are one of the places in the world where they occur in any quantity. Your instructor may provide examples of _dynamically metamorphosed_ rocks during this exercise. In recent years, mylonites and breccias have gained popularity with structural geologists who study past and present plate motions. These rocks allow interpretation of the associated fault movements.

The focus of this exercise will be two major metamorphic rock groups: _regional metamorphic_ rocks that generally possess a foliated texture and _contact metamorphic_ rocks that generally are not foliated. The traditional classification of metamorphic rocks is based on a two-fold subdivision into foliated and non-foliated textures. Once this distinction is made, subsequent classification is based on grain size for the foliated rocks and mineralogy for the non-foliated ones. Dynamically metamorphic rocks may also be classified under this foliated vs. non foliated scheme. Identification and classification of metamorphic rocks is thus quite simple and straightforward, as you will see during this exercise.

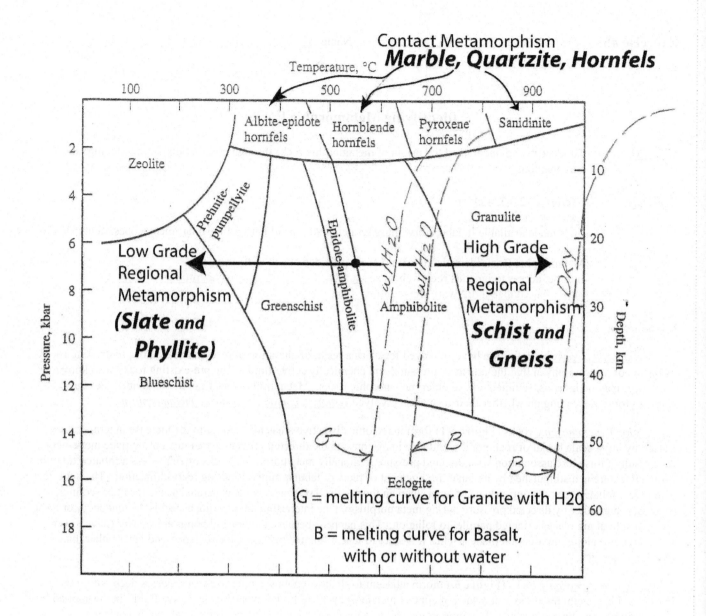

Figure 5.1. Diagram showing pressure and temperature conditions represented by various types of metamorphic rocks studied in this exercise. Melting curves for granite and basalt indicate the temperatures and depths at which these rocks would melt under wet or dry conditions. This illustration is modified from Figure 19.3 in Blatt and Tracy (1996)

Procedure

    Study each of the samples provided by your instructor. Begin by subdividing these into foliated and non-foliated groups. Pay particular attention to the constituent minerals in each rock and its texture as you fill out the *Metamorphic Rock Identification Worksheet*. In addition to naming these rocks, to fully interpret their history please specify (1) a likely protolith, and (2) typical pressure-temperature conditions that prevailed during metamorphism. Most pertinent information is provided in **Key 5** below.

Exercise #6

Name: _____

Class Number: _____

Ordering Geologic Events: Laws of Stratigraphy and Crosscutting Relationships

Objective: To learn about the principles for relative age dating and the rules for understanding the sequences of geologic events.

Materials: textbook
lecture notes
pencil

Introduction

A fundamental aspect of any Earth science research involves the unraveling of events which have occurred in the geologic past. While in many instances the sequence of geologic events in a study area may be quite complex and require extensive field research, the basic interpretation involves only a few fundamental principles. Armed with these basic concepts and a little imagination, even the most complex geologic history can be reconstructed.

To reconstruct geologic history, it is often necessary to rely on two different time frameworks, one _absolute_ and the other _relative_. The absolute age of a rock unit or geologic event is expressed in years from the present; for example, "this rock is 3.5 billion years old." Absolute ages are determined by dating earth materials which undergo radioactive decay. Because the absolute age determination requires sophisticated and expensive laboratory procedures, earth scientists more often rely on relative ages. Dating events in terms of relative geologic time simply means determining whether one event occurred before or after another. In this exercise, we will be dealing exclusively with _relative age determinations_.

Fundamental Relative Age Dating Principles

There are five fundamental principles that need to be considered when determining relative age dates. These will be reviewed below.

**1.** Uniformitarianism   While not directly employed to determine a relative age date, the principle of uniformitariansim nevertheless forms the cornerstone of geologic age dating. This principle was originally conceived by James Hutton in 1795 and popularized by Charles Lyell three decades later; it states quite simply that "the present is the key to the past". In other words, geologic processes that shape the Earth's surface today have likewise shaped the Earth throughout its history.

While taken largely for granted today, the significance of this principle to the nineteenth century scientist was overwhelming. Suddenly the Earth's history could be unraveled. Examples of its application are too numerous to be considered in detail but one significant application is worth noting. It involved early 19th century studies of active glaciers in Greenland and the Alps. Scientists observed that peculiar and distinct deposits form as a result of active glaciation. Scientists looking at ancient rocks noted these same types of deposits, and as a result, they worked out a sequence of numerous "Great Ice Ages" in the recent geologic past.

**2.** Original Horizontality   As first conceived by Nicholas Steno during the late 18th century, the Law of Original Horizontality stated that all sedimentary rocks and most lava flows form in originally horizontal layers. Steno based his principles on observations of sediments being deposited in lakes. He noted that sediments are, deposited under the influence of gravity, and when 30 hey are dumped into standing bodies of water such as lakes, the particles settled out as thin layers parallel to the earth's surface. In general all sedimentary rocks obey this basic law. Carrying his reasoning a step farther, Steno stated that any rocks no longer in a horizontal position, such as those at the very bottom of the Grand Canyon, must have undergone subsequent tilting or folding after deposition. One exception to the Law of Horizontality is the existence of silicic lava flows on modern stratovolcanoes (see Exercise 16) that were originally deposited on slopes as great as 30 degrees.

**3.** Superposition   The Law of Superposition is a corollary to original horizontality. Steno proposed that in any sequence of layered rocks, undisturbed by tilting or folding, the oldest sedimentary layer would be at the bottom of the sequence and the youngest at the top. While originally proposed only for horizontal sequences of sedimentary rocks, the rule can also be applied to folded and tilted strata, so long as they have not been overturned.

4. *Cross-Cutting Relationships* This principle states that any rock unit, fault, or unconformity that cuts across other rock units is younger than the rock units through which it cuts. A few examples serve to illustrate this principle (**Figure 6.1**). In example A, the fault is clearly younger than layers 1, 2 and 3 because it cuts across (displaces) those layers, but it is older than layers 4 and 5 which cross-cut it. In example B, the intrusion must be younger than layers 1-4 because it cuts across all layers. But what about **Figure 6.1C**? At first glance, we might say the intrusion is younger than 1 and 2 but older than 3 and 4 because they appear to cut it, but we must consider another interpretation. This is that layers 1-4 were deposited and then the intrusion occurred, but it did not penetrate layers 3 and 4. Which is correct? In this case, either interpretation may be valid. If we were dealing with a real situation, field work might help us resolve the dilemma.

5. *Faunal Succession* William Smith, an early 19th century engineer, carefully scrutinized fossils that were embedded in rock layers throughout England. He concluded that each layer of rock contained a distinctive faunal assemblage that differed from that in the rock layers above and beneath it. On the basis of this observation, Smith proposed that plant and animal fossils succeed one another in a recognizable order and consequently, certain distinctive fossil groups can be used to correlate relative ages of dissimilar sedimentary rocks from one geographic area to another. An illuminating bibliography of William Smith is popularized by Simon Winchester (2001) in his book: "The Map That Changed the World."

Unconformities

You now have most of the tools necessary to reconstruct geologic events. However, a word needs to be said about unconformities. An *unconformity* is defined as an erosional surface that separates younger layers from older ones. It implies a period of exposure to surface weathering conditions (erosion). Also implied is some geologic event that caused uplift of the rocks to expose them to erosion, usually tilting, folding, or faulting. Unconformities are denoted by wavy lines on the cross sections you will be examining (see example, **Figure 6.2**).

Briefly, unconformities are classified as *angular unconformities*, *nonconformities*, and *disconformities*. These are defined in your text. It might be useful to familiarize yourself with each type of unconformity, although you will not be asked to recognize the types for this exercise, only that an unconformity is present and that it signified uplift and erosion.

Faults and Folds

Earth's lithosphere is broken into plates along fractures known as faults (see also **Exercise #18**). A fault may be defined as a fracture that truncates or displaces distinct geologic units. Because faults are crosscutting features, they are younger than the youngest rock unit in the adjacent rock sequence. Similarly, rock layers that are folded into anticlines or synclines record a compression event that postdates the affected strata. Commonly this compression is caused by plate convergence.

Figure 6.1. Simple cross sections of geologic strata. The sequence of geologic events recorded in each section is summarized below:

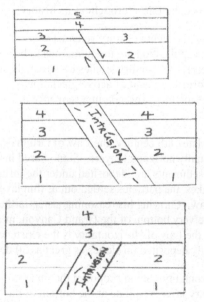

A. Layers 1-3 deposited. Faulting, then erosion (note unconformity). Layers 4-5 deposited.

B. Layers 1-4 deposited, followed by an igneous intrusion (note dike that cuts across layered strata).

C. Layers 1-2 deposited, then crosscut by an igneous intrusion (dike). After a period of uplift and erosion, layers 3 - 4 deposited.

Example of Ordering a Sequence of Geologic Events

To the right of the following geologic cross section (**Figure 6.2**), the sequence of events has been listed in the correct order of occurrence. The oldest geologic event has been placed at the top and the youngest geologic event at the bottom. You will complete similar descriptions for the cross sections on the next few pages.

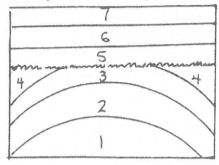

4. Submergence and deposition of units 5, 6 and 7.

3. Uplift and exposure to surface erosion, as denoted by the unconformity (shown by wavy line).

2. Regional compressional event, resulting in folded strata (note anticline).

1. Deposition of units 1, 2, 3, and 4.

Figure 6.2. Example of relative-age determination for geologic events.

Problems

Now is your opportunity to practice the techniques described above. For each of the cross sections below (**Figure 6.3, diagrams A-G**), list the proper sequence of geologic events in the blanks provided. Position the *oldest event on the bottom*; the *youngest on top*. Cross-sections showing strata with various geologic histories. When interpreting the history presented by these sketches, start with event **#1** at the bottom as the first and oldest.

Figure 6.3. Cross sections showing rock strata with various geologic histories.

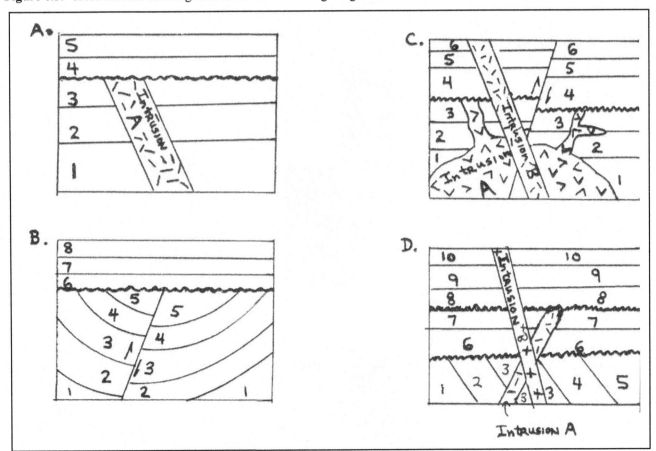

29

E.

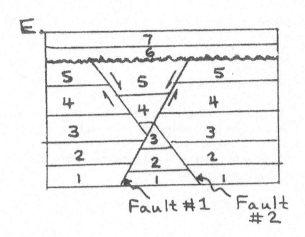

Fault #1   Fault #2

F.

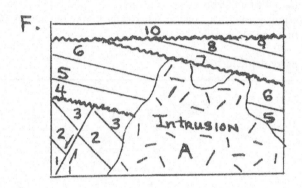

Intrusion A

G.

Lava Flow connected to B

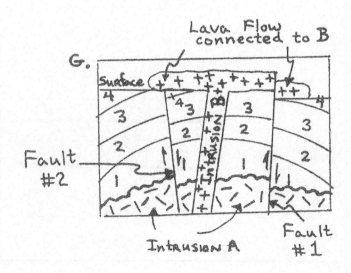

Fault #2

Intrusion A   Fault #1

30

Fossil Preservation

Objectives: To recognize various modes of fossil preservation.

Materials: pencil
hand lens
textbook

Introduction

Much of what we know about the history of life on Earth is derived from a study of _fossils._ Fossils are the remains and/or traces of life forms of former geologic age. Generally, former geologic age means occurring before the Holocene Epoch, the beginning of geologically recent time, about 11,000 years ago. This, in turn, means that objects associated with all ancient civilizations cannot be classified as fossils even though they were excavated from historically old sedimentary deposits. Typically, stone artifacts such as arrowheads of Native Americans or stone tools of aboriginal peoples are not considered to be fossils.

Geologists generally classify fossils into two major groups: body fossils and trace fossils. Body fossils consist of body parts such as shells, bones, teeth, or entire skeletons (of either vertebrate or invertebrates). An ancient clam shell or a dinosaur tooth are examples of body fossils. Trace fossils consist of tracks, trails, burrows, nests, or resting spots of organisms. Much information about behavior of ancient animals has been derived from a study of trace fossils. While even the traces of tree branches that scratched cross mud may be considered as trace fossils, lithified mudcracks, raindrop impressions, ripple marks, or salt crystal casts cannot be considered as fossils because of their inorganic origin.

Fossil Preservation

You are instructed to read the section on fossils and their preservation in your text. You should understand the processes of preservation that occur once organic remains are buried in sediments. Some of the more important preservation modes include:

1. _Original Hard Parts_ (Unaltered Organic Tissue) – here the skeletal parts are not altered chemically or mineralogically during or after burial. The original skeletal tissue is preserved. Bones of mammals from the La Brea Tar Pits are examples.

2. _Permineralization –_ while original organic tissue may remain (as in the cellular part of a bone), open spaces may be invaded and filled with mineral deposits. Dinosaur bones commonly are permineralized.

3. _Replacement –_ this preservation mode sometimes is considered as an extreme case of perminalization in which minerals such as quartz, calcite, pyrite, or hematite almost completely replace the original organic matter in a "molecule by molecule" replacement process. A typical example of replacement is seen in the silicified wood from the Petrified Forest of Arizona.

4. _Molds and Casts_ – when a bone, shell, or other body part is buried and dissolved by ground water an impression (mold) is formed in the rock. This mold later may be filled with subsequent sediment and a cast will result. Many marine mollusks (clams, snails) are preserved this way.

5. _Carbonization (distillation)_ – Under special circumstances of a low oxygen bottom environment, organic remains may becomes buried, compacted, and naturally distilled leaving behind a carbon film. These films commonly show considerable fine structure of the original organism.

6. _Trace Fossils –_ Again, trace fossils commonly take the form of shallow impression (shallow molds and casts), but they also may occur as mineralized droppings (fossilized feces) called coprolites or as stomach stones of certain dinosaurs _(gastroliths)_.

Procedure

Your instructor has set out a number of fossil specimens for you to examine. Your job is to determine:

(a) whether the fossil is a body fossil or a trace fossil and
(b) the dominant mode of preservation illustrated by the fossil specimen. (Some fossils may show a mixture of preservation modes.)

Use the remainder of this page to construct an answer sheet.

Exercise #8

Name: _____

Section: _____

Petroleum Geology

Objectives: To learn some common geologic mechanisms for the generation and accumulation of petroleum.

Materials: pencil
hand lens
knowledge of contouring procedures (Exercise #8)

Introduction

Coal, oil, and natural gas, the so-called fossil fuels, are the fundamental energy sources on which contemporary civilization depends. The "fossil fuels" are so named because they are extracted from the Earth, almost exclusively from layers of sedimentary rock, and have been derived from the chemical transformation of the organic residues of ancient plants (primarily) and animals. In this exercise, we will ignore coal (a solid fossil fuel) and concentrate on the liquid and gaseous fossil fuels which we called *hydrocarbons.*

Oil and gas technically are referred to as hydrocarbons because hydrogen and carbon are dominant elements in their chemical structures. Typically, straight chains, branched chains, or rings of carbon atoms are linked with hydrogen atoms. Other chemical elements such as sulfur, oxygen, nitrogen, and even some metals may be attached at specific sites to the basic hydrocarbon molecule. Crude oil may contain a wide range of hydrocarbon structures of variable chemical composition.

Although petroleum (oil; originally called "rock oil") likely has been a common component of sedimentary rocks since abundant life forms evolved on Earth, a rather limited set of conditions is necessary to concentrate the hydrocarbons into locations and accumulations of such volume as to guarantee a commercial deposit when discovered and tapped by humans. Some of the more important (but by no means all) of the prerequisites for a money-making petroleum accumulation are:

1. A rich supply of *organic material,* ideally marine algal cells that accumulated in oxygen poor bottom sediments sometime in the geologic past. This situation will result in a hydrocarbon-rich *source sediment* and ultimately a rich *source rock* in which oil and gas are generated from the organic residues. A typical example of a source rock would be a black shale or dark fetid limestone.

2. A *heat source*, not too hot (but not too cool, either), in the general range 65 - 125° Celsius is required to break down and reform the organic residues into hydrocarbons.

3. In order for oil and gas to be transmitted and stored in a rock body, that rock must be both porous and permeable. (NOTE: the student is advised to review and understand these terms at this point. If they are not covered in the text, the instructor will explain them. A porous and permeable rock that allows for migration and accumulation of fluids is called a *reservoir rock*; one example would be a well-sorted, weakly cemented, clean sandstone. Hydrogeologists refer to such rocks as *aquifers*.

4. Before a commercial accumulation of fluids can form, the hydrocarbons must be localized by a *trap*. Generally, traps consist of geologic *structures* that limit the upward (or lateral) migration of fluids by bounding them with a rock type of a lower porosity and permeability. An anticline in a sandstone/shale sequence would be one example of a *structural trap*. A fault that juxtaposes two rock types of different porosity and permeability values is another example. Some traps involve neither folding (as in anticlines) nor faulting, but simply a change in the rock type from one degree of porosity/permeability to another. This is termed a "facies change" and results in a *stratigraphic trap*. A "*stratigraphic pinch-out,*" where sandstone grades into shale, as along the shore of an ancient sea, is one example of a stratigraphic trap.

5. A trap should be covered with a fine-grained. Impermeable rock called a *cap rock* to prevent leakage of fluids from the reservoir. (The *reservoir* consists of the porous and permeable rocks enclosed in the trap.) Common cap rocks include shale or any *evaporite*.

6. Timing is crucial! Economically viable petroleum targets require an ideal geologic history. The reservoir and trap must remain buried or at least not exposed to erosion which would cut it and release fluids to the surface. Also, subsequent to its formation, the trap must not become so deeply buried that geothermal processes break down (degrade) the hydrocarbons.

Procedures: Problem One

Your instructor has placed various samples of sedimentary rock at the front of the room. From these displayed specimens, please select a single specimen that would best qualify as an example of a good _source rock_ for oil and/or gas. Use criteria of color, grain size, and any special features of the rock in your evaluation. Please **defend** your selection in the space provided below:

Similarly, a set of potential reservoir rocks lies been supplied to you. Using criteria such as mineralogy, grain size. sorting, and apparent porosity, select a specimen that you feel is the best example of a _reservoir rock_ and _defend_ your choice in the space provided below:

Procedures: Problem Two

Petroleum geologists spend much of their time preparing subsurface maps in their search for buried traps. One widely used type of subsurface map is the _structure contour map._ Typically, the structure contour map is used to show the high and low points of a buried structural feature (such as an anticline). The projection can be rotated to show a three-dimensional view. Contours are drawn (just as in topographic contouring) on top of a buried rock unit of interest such as an oil soaked reservoir rock or other unique rock layer which serves as a marker bed. Information on the depth of the marker bed comes from drill hole data or seismic reflection horizons.

At this point, it would be good to review the rules of contouring that we use in constructing topographic maps (**Exercise #9**). The principles of contouring are the same! Structure contour maps such as the one used here are drawn on the top of a deeply buried horizon. We simply have far fewer control points!

You are given a base map with the plotted location of 16 drill holes (**Figure 8.1**). Each hole was drilled to the _top_ of a significant oil sand although each borehole did not necessarily result in a commercial oil well. Your job is to draw a structure contour map on top of this buried oil sand using the sub-sea data points given below for the 16 previously drilled wells. Use a contour interval of 100 feet. Next, _name_ (identify) the specific type of hydrocarbon trap you have defined by contouring this buried feature. Finally, show _on the map_ where you, as a petroleum geologist/engineer, would locate the next well (#17).

Well Data (Modified from Poort, 1980, Historical Geology: Interpretations and Applications):

Well Number	Depth to Top of Formation (Subsea)
I	2535
2	2720
3	2520
4	2738
5	2744
6	2854
7	2836
8	2612
9	2596
10	2729
11	2854
12	2540
13	2941
14	2556
15	2964
16	2722

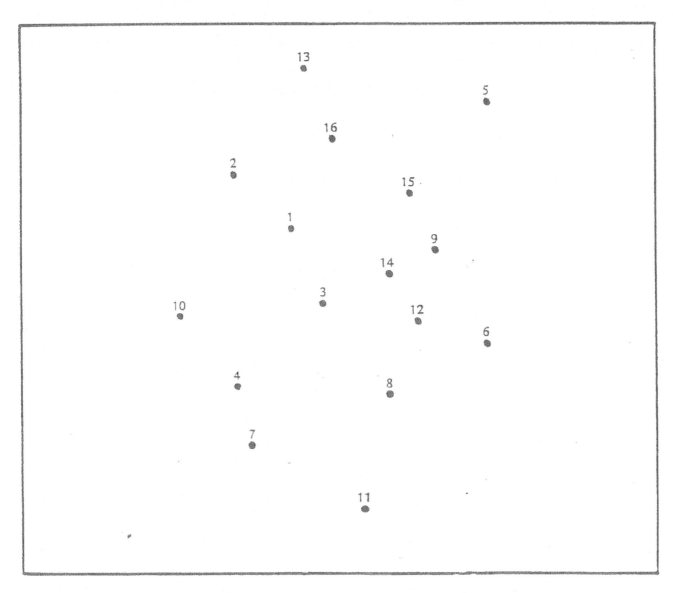

Figure 8.1. Structure contour map base for Problem 2. Use a 100 foot contour interval.

Procedures: Problem Three

Please study the following *geologic cross section* (**Figure 8.2**; kindly provided by Dr. J. Nourse). Please decide, based on the geologic data available, exactly where to drill for hydrocarbons. Unfortunately, your drilling budget allows for <u>only four</u> wells! Consequently, you must select your drilling locations carefully. In the expensive and highly competitive world of oil exploration "close enough" doesn't count.

After selecting your drill site locations, extend <u>vertical</u> lines (your drill pipe) from the surface down to the "oil pool" in the reservoir rock. Again, use vertical drill holes only. No directional or horizontal drilling will be employed here.

Next, indicate by shading (on the cross section) exactly where you expect the hydrocarbon accumulation to occur. Also, show with arrows on the diagram the direction of fluid movement (*migration*) into your wells.

Finally, *number* each specific type of trap on the diagram and name it in the space provided below. Also describe the source rock and the reservoir rock. We hope that you will enjoy this exercise because you are now a PETROLEUM GEOLOGIST for a day!

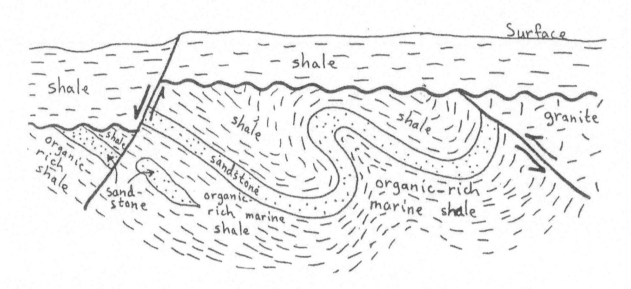

Figure 8.2.2. Geologic cross section for Part Three. Drawing by J. Nourse.

Please describe your petroleum targets below:

Exercise #9

Name_____

Section_____

Interpretation of Topographic Quadrangle Maps

Objective: To understand how to read a topographic map, and to use topographic maps to interpret changes through time recorded in the area surrounding the Cal Poly Pomona campus. *Note*: This exercise was specially designed by the authors to educate Cal Poly students about the historical development of the Pomona-San Dimas-Diamond Bar-West Covina area.

Materials: ruler
calculator
topographic maps (available in lab)
pencil
working knowledge of topographic map principles (Exercise #7)

Introduction

In this exercise, we will be using three versions of the San Dimas 7.5' quadrangle topographic map. Topographic maps give information about the elevation of the land surfaces shown on the map; a 7.5' quadrangle (or quad) indicates that the map covers an area 7.5 minutes of latitude by 7.5 minutes of longitude. If your understanding of latitude and longitude and their subdivisions is hazy, be sure to read about them in your text before you come to this recitation.

The first group of questions is based on the 1966 version of the San Dimas, California, 7.5' topographic quadrangle, and the second part (Problems 7-8) utilizes the 1954 version. Please note that the number of copies of the older map is limited, so when you have the chance to work on that map, you should do so. Problem 9 is best addressed by observing the 1981 photo revision of the 1966 map. Anything shown in purple on this map was added between 1966 and 1981. To address Problem 10, students might look at Google Earth to compare present-day features to a hard copy of the 1981 photo-revised map.

Problems

1. What is the approximate area of the map in square miles? What is the area in square kilometers? (Remember that 1 mile = 1.62km.) Please show your work below:

2. Declination is the angular difference between the direction to magnetic north (m. n.) and true north (indicated at the bottom of the map by a "North Star"). Remember the declination changes gradually with time.

What is the declination between true north and magnetic north indicated on the <u>1966</u> map? Express your answer in degrees.

What is the declination on the 1981 photo-revised map?

What is the declination on the 1954 map?

3. A major type of information on a topographic map is elevation, which is shown with brown contour lines that connect points of equal elevation. Refer to Exercise #8 if you need more details about reading and interpreting the contour lines. . What is the approximate elevation of the old Administration Building (Building 1) at Cal Poly Pomona? This is the "L" shaped building with the flag. Note that this building rests on a slope, so express your answer as a <u>range</u> of elevations.

4. Where is the highest point on the map (use compass for directions: for example, SW corner of the map, or just N of a particular intersection)? What is its elevation (to the nearest 20-foot contour)? The map will not necessarily indicate what the highest elevation is: you just have to look for it. Use what you know about the regional topography (i.e., where are the closest mountains?) to make an educated guess about where to start your search.

5. What is the elevation of the lowest point on the map and where is it on the map? Hint: water tends to accumulate at low elevations as dictated by the law of gravity.

6. What is the total relief on this map? (Relief is the elevation *difference* between the highest and lowest points on the map).

7. On the 1954 version of the San Dimas quadrangle, what was the type of land use in the area immediately to the southeast of where the Cal Poly Pomona campus is today? Hint: the green dots are arranged in regular rows and columns.

8. List three notable changes that occurred in the area between 1954 and 1966 (look for differences between the two maps). "Urbanization" is too general; try to find specific examples. Remember that you cannot say anything about changes in vegetation between 1954 and 1966, because the 1966 version of the map simply doesn't contain any information about vegetation on it (the green overprint is optional).

a. _____

_____

b. _____

_____

c. _____

_____

9. List three notable changes that occurred in the area between 1966 and 1981 (look for differences between the two maps):

a. _____

_____

b. _____

_____

c. _____

_____

10. Describe at least two changes that have occurred in this area since 1981 (i.e., is there anything that exists today but is not shown on the 1981 revision?). Again, be specific.

a. _____

_____

b. _____

_____

Construction of Topographic Map and Profile

Objective: To learn how to construct a topographic map and how to draw a profile from a contour map.

Materials: ruler
eraser
pencil

Introduction

In **Part l** of this exercise, you will prepare your own topographic map so that you can understand the basic fundamentals of mapping. Although all map preparation today is automated, the same basic principles still apply. *Topographic maps* show the surface of the Earth as a series of lines called contours. A *contour* is defined as a curve that connects points of equal elevation. *Elevation* is defined as the distance above sea level, usually measured in feet or meters.

Topographic contour maps for most areas of the United States have been prepared by the United States Geological Survey. You might ask yourself who besides a geologist would want to use them. The answer is that most topographic maps are purchased by building contractors and civil engineers and others involved in construction. A significant number of topographic maps are also sold to hikers and backpackers, because these maps often show trails and other information not found on standard highway maps.

As stated above, contour maps show the Earth's surface as a series of contours, with each separate contour representing a particular elevation. By convention, sea level is taken as the 0-foot contour. The vertical distance between contours, the contour interval, varies from map to map and is a function of the steepness of the local topography. For instance, a map of a portion of the San Gabriel Mountains might have a 40-foot contour interval because the topography is very rugged, and a 5- or -10 foot contour interval wouldn't be able to show the topography clearly because the contours would be too closely spaced. In contrast, a topographic map of a portion of Kansas might use a 5-foot contour interval because a map with a 100-foot contour interval might have only one line on it! You will also find that every fifth contour line on a topographic map is darkened and labeled with the elevation of that contour line. As an example, if we are using a 100-foot contour interval and our first contour is 0 feet, we would have contours at 100, 200, 300, 400, and 500 ft., with those at 0, 500, 1000 ft, etc. darkened and labeled.

In **Part 2** of this exercise, you will construct a topographic profile. A topographic profile is a diagram that shows the change in elevation of the land surface along any given line. It represents graphically the "skyline" as viewed from the distance. Features shown in profile are viewed along a horizontal line of sight while those shown on a map are viewed from the distance. Features shown in profile are viewed along a horizontal line of sight while those shown on a map are viewed along a vertical line of sight (i.e., looking down). Topographic profiles can be constructed along any given line.

The vertical scale of a profile is arbitrarily selected and may be larger than the horizontal scale of the map from which the profile is taken. When such is the case, we say that the profile has vertical exaggeration. As an example, a profile with a horizontal scale of 1" = 1000' and a vertical scale of 1" = 100' has a vertical exaggeration of 10 times.

Procedure / Problems

Part 1 – Topographic Map

So just how do you go about making a topographic map? First, take a look at the example map given in **Figure 10.1**. This is what a hand-drawn topographic map looks like. The small dots with numbers beside them represent the elevation of the land surface at that point. These so-called "spot elevation" are what you use to determine where to draw the contour lines. The contour interval is 20 feet, so the contour lines have values of 500, 520, 540 ft, etc.

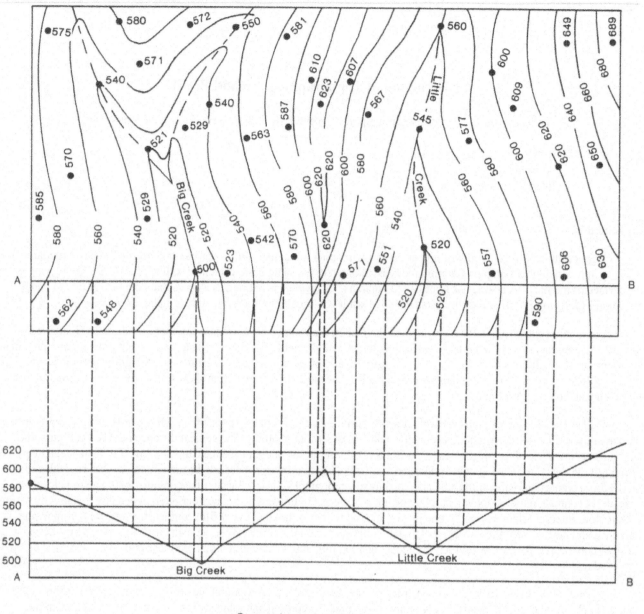

Figure 10.1 Sample Topographic Map, constructed from spot elevation, and a sample profile.

The most difficult thing about drawing a topographic map is figuring out where to start. Generally, the best way is to find the highest or lowest spot elevation on the map and begin there. As an example, let's say the lowest point of a map is at 515 feet. Because the contours will all have values that are multiples of 20 (the contour interval you will be using), the first line you will draw is the 520-foot contour line. Then look at an adjacent spot elevation. Let's say there is one nearby that has an elevation of 525 feet. Because 520 lies between 515 and 525, the 520-foot contour line must lie somewhere between these two points, in this case, exactly halfway between them. Put a light mark there. If the points had values of 515 and 538 respectively, the same contour line would lie closer to the 515 point (see **Figure 10.2**). For this exercise, we won't be all that concerned about accuracy so long as you get the contour lines between the proper spot elevations. Examine all the points adjacent to the 515 point in turn, decide where the 520-foot contour line might lie between them and put a light mark there if it does. Then draw a smooth line through the marks, and you have completed your first contour line!

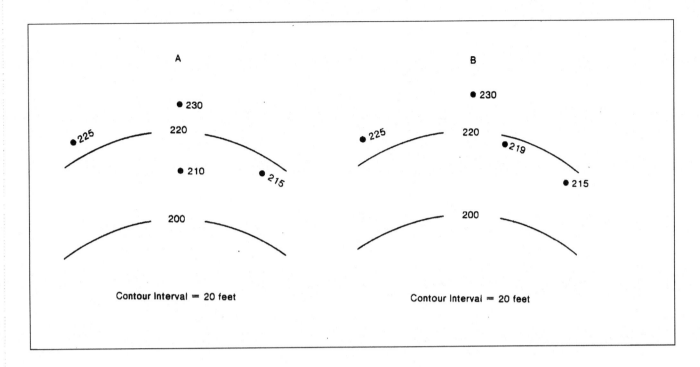

Figure 10.2. Two examples of contouring using spot elevations. In diagram A, the 220-foot contour passes midway between 210 and 230 ft; in B, the 220-foot contour passes close to 219 ft.

Now that you know how to draw a contour line, there are a few contouring rules you will have to follow as you draw your topographic map:

1. Contour lines are smooth, curving lines. Don't just connect cots. When finished, your map should have a regular pattern (see **Figure 10.1**).

2. Contour lines never intersect or cross.

3. Closed contours, appearing on the map as ellipses or circles, almost always represent hills or knobs. (In the rare instances where the closed contours indicate a circular depression or a crater, the contour lines will sometimes have small marks pointing into the center of the depression.)

4. Steep slopes are shown by closely spaced contours; gentle slopes by widely spaced contours.

5. When contour lines cross streams, they bed upstream; that is, the segment of the contour line near the stream forms a "V" with the apex of the V pointing upstream toward the higher elevation. If this doesn't make sense, think about what happens when you cross a stream. From the bank, you have to step down into the water, which means the bottom of the stream is at a lower elevation than the banks. So, if we say a contour line connects points of equal elevation, then it makes sense that to find a point on the stream bed that has the same elevation as the two banks of the stream, we are going to have to go upstream (see **Figure 10.3**).

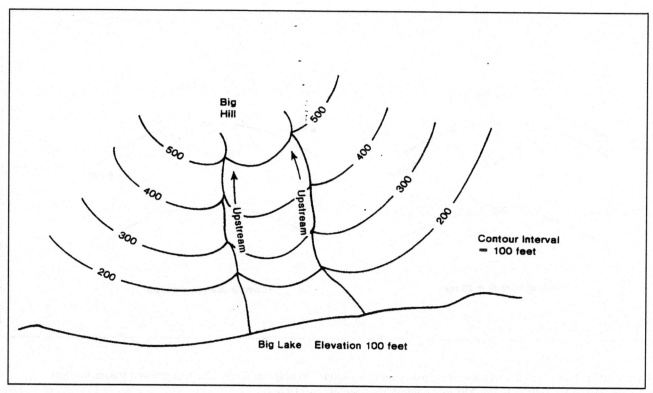

Figure 10.3. Simple contour map showing topographic "V's" with the apex pointing upstream.

The map you will be contouring is **Figure 10.4**. Use a 20-foot contour interval and be sure to follow the contouring rules!

Part 2 - Topographic Profile

This part of the exercise involves the construction of a topographic profile from the contour map you prepared in Part 1. Below are the steps for drawing a topographic profile. Refer to **Figure 10.1** for help if necessary. You will be constructing this topographic profile for the contour map that you prepared in Part 1 of this exercise, so you should follow the instructions below for your profile on **Figure 10.4**.

1. The line along which your profile will be constructed is **A-B**. Draw this line on your contour map (from A to B) to help you.

2. On the topographic map, note where each contour line crosses your profile line A-B. It may help you to mark each of these intersections with a pencil. Note also where streams, lakes, or hilltops lie on your profile.

3. Project the intersection of each contour line with the profile line A-B on the topographic map straight downward to your topographic profile. At a point on the profile equal to the elevation of your particular contour, place a dot. Again, the example in **Figure 10.1** should help you.

4. Connect these dots with a smooth curve and label any significant features such as streams.

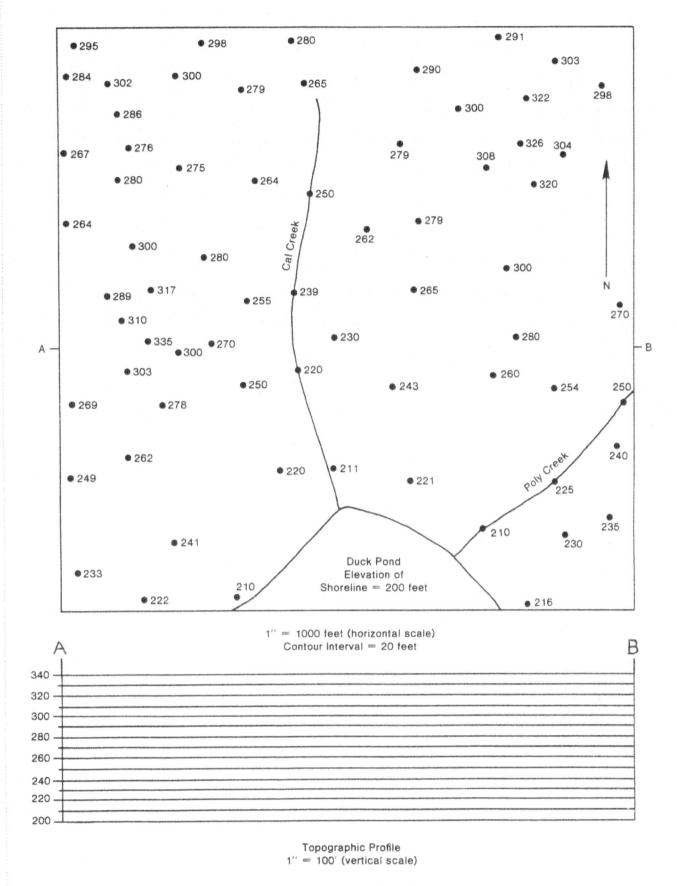

Figure 10.4. Hypothetical topographic map, with topographic profile (worksheet).

45

Name: _____

Section: _____

Volcano Profiles, Shapes, and Volumes

Objectives: Students shall utilize topographic maps and true-scale topographic profiles as a means of visualizing the morphologies of silicic and basaltic volcanoes. Quantitative information from these images will then be applied to slope and volume calculations

Materials: pencil
eraser
colored pencils
ruler
several sheets of graph paper
protractor

Introduction

Active *volcanoes*, interesting to nearby inhabitants for obvious reasons, have long been studied by scientists. This exercise explores commonly used methods for describing the physical shape and size of volcanoes, i.e., volcano *morphology*. Students will study topographic maps of several famous volcanoes and use elevation and scaling information on these maps to draw accurate illustrations. As this may be your first experience with topographic maps, please listen carefully to your instructor as you are guided through the exercise. Also, refer to Exercise #8 for a review of procedures for topographic profile construction.

Procedure / Problems
1. Refer to **Figures 11.1** through **11.4**, study the map scales, and try to visualize the 3-dimensional geometry of the volcanoes depicted. One map illustrates Mount Saint Helens (elevation = 9677ft) before its cataclysmic eruption, the second shows it a few years later after a lava dome had grown within the resulting crater. The map of Crater Lake shows a *caldera* that formed after the eruption of Mt. Mazama in 5677 B.C. Notice the lake that has partly filled up the crater, and the "resurgent lava dome" (Wizard Island) that grew about 1000 A.D. The last map illustrates the big island of Hawaii, which most beginning geology students know is composed of several overlapping basaltic shield volcanoes. For additional information on these and other volcanoes, please refer to the following web sites:

      http://vulcan.wr.usgs.gov
      http://hvo.wr.usgs.gov
      www.volcanolive.com

2 A. Construct a true-scale topographic profile through **Figure 11.1** along the line indicated, using the box provided in **Figure 11.1A**. Notice that the vertical scale is set equal to the map scale, so angles and dimensions measured on your profile will be true. Use each **bold-face** contour crossed by the profile line for elevation control. Be sure to show Pumice Butte.
Quantitative Applications:
B. Use a protractor to measure the maximum and minimum slope angles in **degrees**. Indicate these angles on your profile.
C. Calculate the volume of Mount Saint Helens located above the 4600 ft elevation. The method of horizontal slices would work here; although approximation as a cone will yield faster results. Recall that $V_{cone} = 1/3\pi r^2 h$. Express your answer in **cubic feet**.

3 A. Construct a true-scale topographic profile across the present-day cratered version of Mount Saint Helens (**Figure 11.2**) along the line indicated. Use the box provided in **Figure 11.2A**. You may find it helpful to refer to the photograph of Mount Saint Helens reproduced in **Exercise 16**. Its elevation has been reduced to 8364 ft. Notice also the lava dome, which should show up on your profile.
Quantitative Applications:
B. Compare this profile with that constructed for Problem 2A above. Then calculate total volume of rock that erupted during the 1980. Various methods* are possible; ask your instructor for hints to facilitate your work. Express your answer in **cubic feet**.

*To apply the method of horizontal slices, proceed as follows: Use a red colored pencil or pen to highlight all of the map contours. Then use different colors to fill in the separate areas residing between the 4600 and 5500 ft contours, the 5500 and 6400 ft contours, the 6400 and 7300 ft contours, the 7300 and 8200 ft contours, and the small areas above the 8200 ft contour. These colored regions define horizontal slices, each with a thickness of 900ft. Essentially, you need to figure out the <u>average map area</u> covered by each slice, multiply each average area by 900 ft thickness, then sum the resulting volumes. The difficult part is acquiring the average areas: use graph paper, the map scale, and a "counting squares" approach to determine the area enclosed by the 4600ft, 5500ft, 6400ft, 7300ft, and 8200ft contours that bound the top and bottom of successive slices. The average area of each slice is just the average of its top and bottom areas. Once you have calculated the post-1980 volume of Mt. Saint Helens in cubic feet, subtract this volume from the volume calculated in Problem 2C to determine the total volume of rock erupted.

C. The rock volume calculated above in 3B was pulverized and mixed with air during the explosive eruption, resulting in a blanket of volcanic ash. Assuming that the volcanic ash had a porosity of 65%, calculate the total area (in **square miles**) it would have covered if the deposit were uniformly one foot thick.

4 A. Refer to the topographic map of present-day Crater Lake (**Figure 11.3**). Use a red pen to highlight the bold-face <u>elevation</u> contours crossed by the line of profile. Then use a dark blue pencil or pen to highlight the contours in the lake. Notice that these contours indicate <u>depth</u> (distance in feet measured down from the lake surface). You should be able to define the 0ft, 240ft, 600ft, 1200ft, and 1800ft depth contours. Color in the area covered by the lake with a light blue.
B. Now let's draw a topographic profile, this time applying some vertical exaggeration so that details of the crater can be emphasized. Using the profile box provide **in Figure 11.3A**, carefully construct the profile with the given vertical scale of 1 inch = 2000 ft. Use the bold-face contours on land and all of the contours within the lake as elevation control points. Show both the lake surface and the bottom of the lake on your profile. The lake surface has an elevation of 6176ft.

Quantitative Applications:
C. Calculate the volume of water contained in Crater Lake. Express your answer in **cubic feet**. Again, the method of horizontal slices is probably most accurate.
D. Mount Mazama was originally about 12,000 feet high. How much rock was blasted away during the spectacular eruption in 5677 B.C.? Express your answer in **cubic feet**. *Hint:* Extrapolate the original volcano profile onto the profile you have already drawn, then determine the missing volume assuming that the original volcano shape was a cone.

5 A. Construct a true-scale topographic profile across the Big Island of Hawaii (**Figure 11.4**) along the line indicated. Use the profile box provided below the map. Notice that the scale is very different than that in the previous three problems. At this scale, you will only be able to utilize a few of the contours crossed by the profile line. Construct your profile with a very sharp pencil.
B. It turns out that most of the Hawaiian shield volcano is submerged beneath the ocean. Let's visualize this by drawing the sub-sea topography. Attach extra paper to the left and right sides of your profile of 5A above. Then plot points to show the depth of water (vertical dimension below sea level) at various distances measured out from the shoreline. Here is the pertinent data:
<u>Left side of profile:</u> Depth = 6000ft, 12,000ft, and 15,000ft at distances of 9 miles, 20 miles, and 30 miles, respectively.
<u>Right side of profile:</u> Depth = 6000ft, 12,000ft, and 18,000ft at distances of 12 miles, 18 miles, and 24 miles, respectively.

Quantitative Applications:
C. Use a protractor to measure the maximum and minimum slope angles in **degrees**. Indicate these angles on your profile.
D. How high does the Big Island volcanic complex stand above the ocean floor? Express your answer in **feet**
E. What is the total volume of basalt contained in the Big Island shield volcano complex (above and below sea level)? *Hint:* The fastest way to do this problem is to measure pertinent dimensions (i.e., height and radius) directly off your profile of 5B above, then approximate the volcano shape as a cone. Recall that $V_{cone} = 1/3\pi r^2 h$. Express your answer in **cubic feet**.

6. Contrast the morphology (shape and size) of Mount Saint Helens prior to its 1980 eruption with that of the volcano complex on Big Island of Hawaii. Refer specifically to information you have acquired from your topographic map and profile analysis. You may wish to discuss differences in slope angle, volume, height, width, etc.

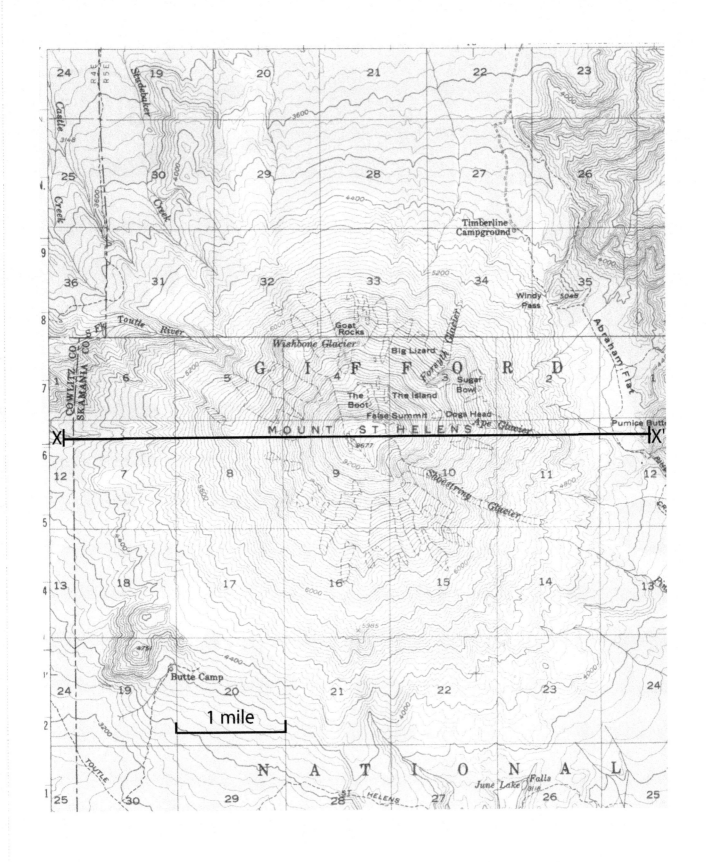

Figure 11.1. Topographic map of part of the Mt. Saint Helens 1958 USGS 15' quadrangle. Contours in feet above sea level.

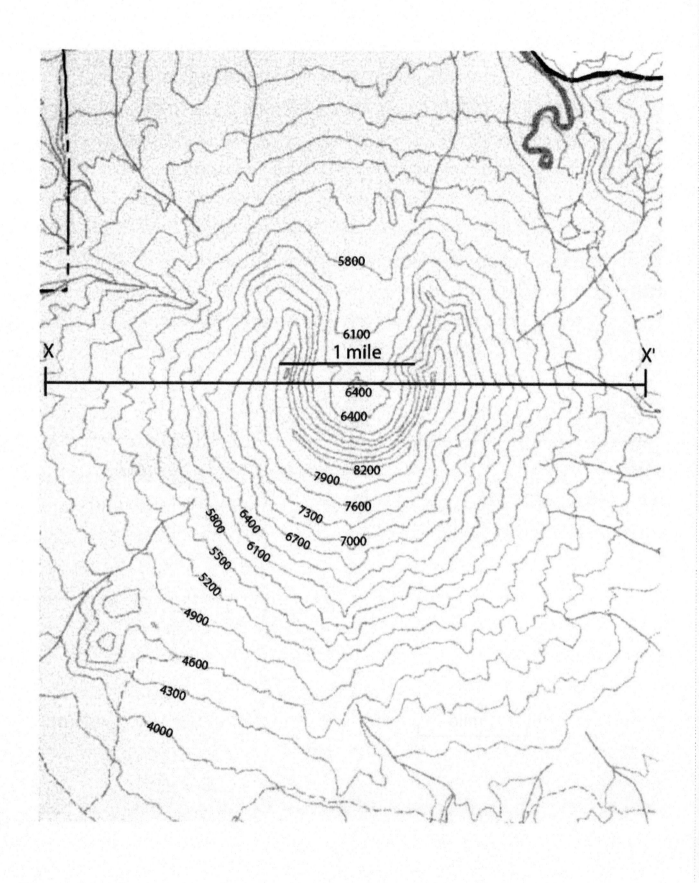

Figure 11.2. Topographic map of Mt. Saint Helens after its 1980 eruption. Contours in feet above sea level.

Figure 11.3. Topographic map of part of the 1958 USGS Crater Lake National Park 15' quadrangle. Contours are in feet.

51

Figure 11.1A. Topographic profile of Mount Saint Helens prior to its 1980 eruption

Figure 11.2A. Topographic profile of Mount Saint Helens following its 1980 eruption

X′

8000 ft
6000 ft
4000 ft
2000 ft
0 ft

X

Figure 11.3A.
Vertically exaggerated
topographic profile through
Mt. Mazama caldera and
Crater Lake, Oregon

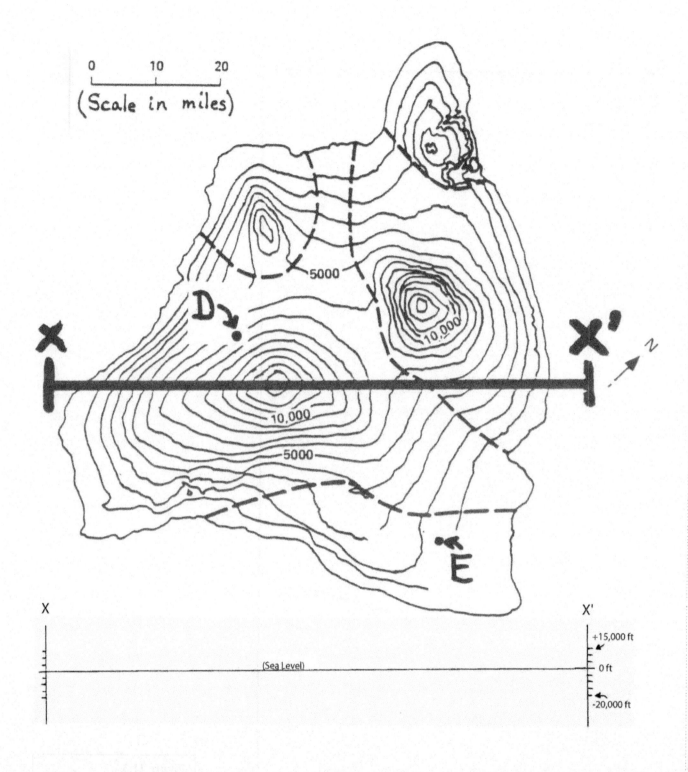

Figure 11.4. Topographic map and true-scale topographic profile of the Big Island of Hawaii. Contours are in feet above sea level.

Exercise #12

Name_____

Section_____

Construction of Geologic Maps and Geologic Cross Sections

Objectives: To acquaint students with valuable three-dimensional information available from typical geologic maps. Methods of constructing geologic maps and cross sections will be introduced through analysis of several hypothetical areas where rock layers have been folded and faulted.

Materials: pencil
colored pencils—blue, brown, red, green
protractor
ruler
calculator

Introduction

Geologic maps and geologic cross sections are a fundamental tools used by geologists to visually communicate the three-dimensional configuration of rocks and soils in a given location. _Geologic maps_ generally show the distribution of various rock and soil types and the orientation of specific structures that intersect the ground surface. Corresponding _geologic cross sections_ provide a profile view of geology. These maps and cross-sections are not just "pretty pictures" that display the outer layers of Earth's crust. They contain valuable quantitative information useful to scientists and engineers, for example, angle of sloping ground, thickness and volume of specific rock units, precise location of buried features, the degree of deformation experienced by rock layers, and the location or orientation of fractures and faults.

The basis for interpreting geologic cross sections from surface geologic mapping is the projection of planar and linear features into the subsurface. Geologists measure such geometric features in the field with a special tool called a "Brunton compass." The drawing below illustrates the concept of "strike" and "dip, " terms that quantitatively describe the orientation of planar features. Geologists use this geometric information to depict the configuration of inclined layers in both map and cross section view. Notice that dip is always perpendicular to strike, and the number gives the angle of "tilt" or dip, in degrees, as measured from a horizontal reference line. Special symbols are used to indicate horizontal layers, vertical layers, and overturned layers. Your instructor will explain these to you.

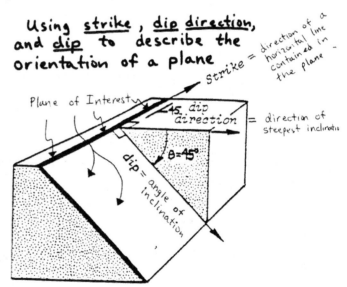

Figure 12.1. Geometric definition of **Strike** and **Dip** for an inclined plane.

55

Procedure (Part 1, A, B, C). Construction of Geologic Maps and Geologic Cross Sections From Raw Field Data:
Your instructor will lead you through the procedures for constructing a geologic map and cross section from a raw data base. *The following steps apply to construction of the geologic maps and cross sections presented in Figures 12.2, 12.3, and 12.4.* Specific data sets (compiled from a geologist's field notes) are listed after the general instructions. Your instructor will assign you one or more of these maps. Please use a sharp pencil for your initial construction. You may add color and ink later if you wish to enhance your visual presentation:

A. Use a protractor to plot a strike and dip symbol at each station on the map where this information was measured. Write in the specific value of dip.
B. Plot the lithology near each observation station, using standard geologic patterns. For shales or siltstone, make "grain" of pattern parallel to the strike bar to enhance the orientation of layering.
C. Sketch in lithologic contacts (boundaries between layers). Project these contacts parallel to strike (i.e., parallel to the long bar of the strike/dip symbol).
D. Mark all places on the map where the cross section line intersects lithologic contacts.
E. Project these contact locations vertically down to the topographic surface on the cross section. Then use a protractor to project each contact into the subsurface.
F. Fill in appropriate patterns to distinguish each rock unit on the cross section.
G. Indicate axes of anticlines and synclines on the map; show direction of plunge if appropriate.
H. Construct a simple legend that explains patterns corresponding to lithologic units and all other map symbols used. Organize rock units so that <u>oldest layer is on the bottom, youngest on top</u>.
I. Write in specific layer thickness on your legend for those units that thickness can be measured.

Field Observations (Raw Data) for Figure 12.2:

Location:	Lithology (Rock Type)	Strike/Dip of Sedimentary Layering:
A	siltstone	N4W/25SW
B	conglomerate	N9W/27SW
C	granite, no layering	
D	siltstone	N15E/59SE
E	sandstone	N25E/55SE
F	conglomerate	N12E/60SE
G	granite, no layering	
H	conglomerate	N17W/30SW
I	sandstone	N10W/28SW
J	siltstone	N19W/30SW
K	granite, no layering	
L	siltstone	N24E/62SE
M	sandstone	N30E/55SE
N	conglomerate	N86E/15SE
O	sandstone	N78W/12SW

Field Observations (Raw Data) for Figure 12.3:

Location	Lithology	Strike/Dip of Sedimentary Layering
Site A	siltstone	N10E/68SE
Site B	sandstone with pebble conglomerate	N10E/65SE
Site C	shale	N15E/50SE
Site D	sandstone with pebble conglomerate	N70E/12SE
Site E	siltstone	N45W/25SW
Site F	sandstone with pebble conglomerate	N45W/25SW
Site G	shale	N40W/22SW
Site H	sandstone with pebble conglomerate	N12E/70SE

Field Observations (Raw Data) for Figure 12.4:

Location:	Lithology (rock type)	Strike/Dip of Sedimentary Layering:
a	siltstone	N20E/40NW
b	sandstone	N20E/40NW
c	conglomerate	N20E/40NW
d, e, f, q, r	granite, no layering	
g	conglomerate	N10W/75NE
h	sandstone	N10W/75NE
i	siltstone	N10W/75NE
j	shale	N30W/60NE
k	shale	N50E/15NW
l	siltstone	N40E/15NW
m	sandstone	N40E/15NW
n	sandstone	N75E/20NW
o	siltstone	N90E/35N
p	conglomerate	N30W/50NE
s	conglomerate	N10E/40NW
t	siltstone	N65E/35NW
u	sandstone	N5W/70NE
v	shale	N/50E
w	shale	N15E/20NW
x	conglomerate	N80E/30NW

Procedure (Part 2). Construction of Geologic Cross Sections Given A Completed Geologic Map:
Now look carefully at the various geologic maps provided in **Figures 12.5- 12.10**. Notice that the cross section boxes have not been filled in. Your instructor will assign one or more of these for you to complete as homework. These also make great study material for examinations! For each geologic map, study the strike and dip symbols and try to visualize the geometry of the inclined layers. Most of these problems involve at least one fault plane. You can learn more about the geometry of faulting in Exercise #19. The basic procedure for each assigned cross section is similar. *Please follow the steps below*:

A. First, draw in symbols for synclines and/or anticlines on this map where appropriate.
B. Locate the fault on the cross section and project it to depth at the angle specified for the dip.
C. Construct a cross section through the line indicated, and complete the legend, following steps **1D** through **1I** above. The layers are folded so try to connect similar rock types at depth. Take care to preserve a uniform thickness for each layer. Assume that the fault is active; therefore it should cut across all layers.

Congratulations!! You have now developed the fundamental skills necessary to visualize and interpret much of the geologic data presented in the succeeding Exercises.

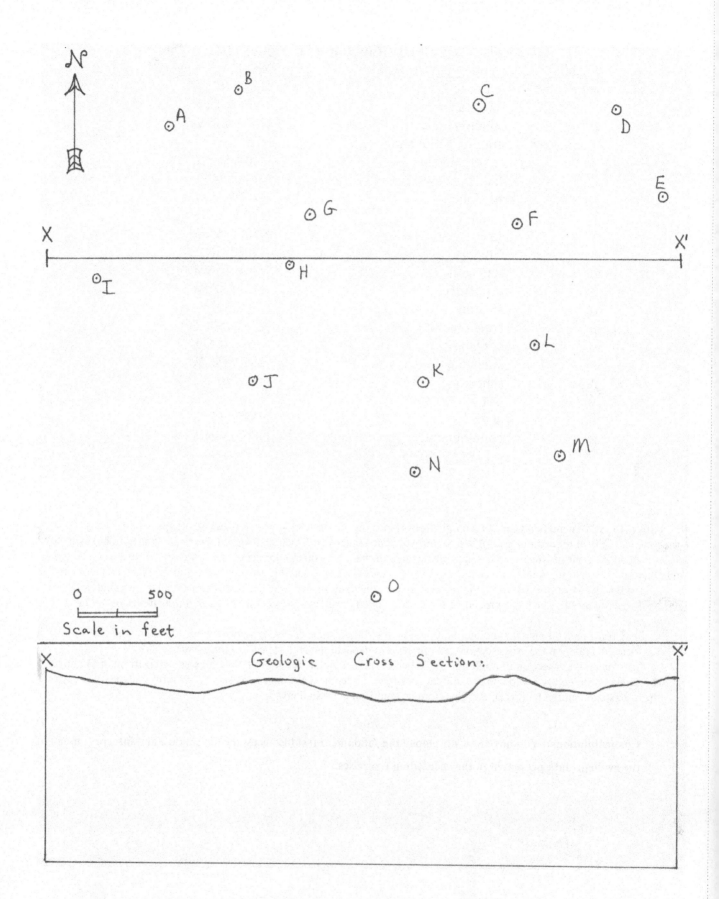

Figure 12.2. Base map for construction of Geologic Map and Geologic Cross Section (**Part 1A**).

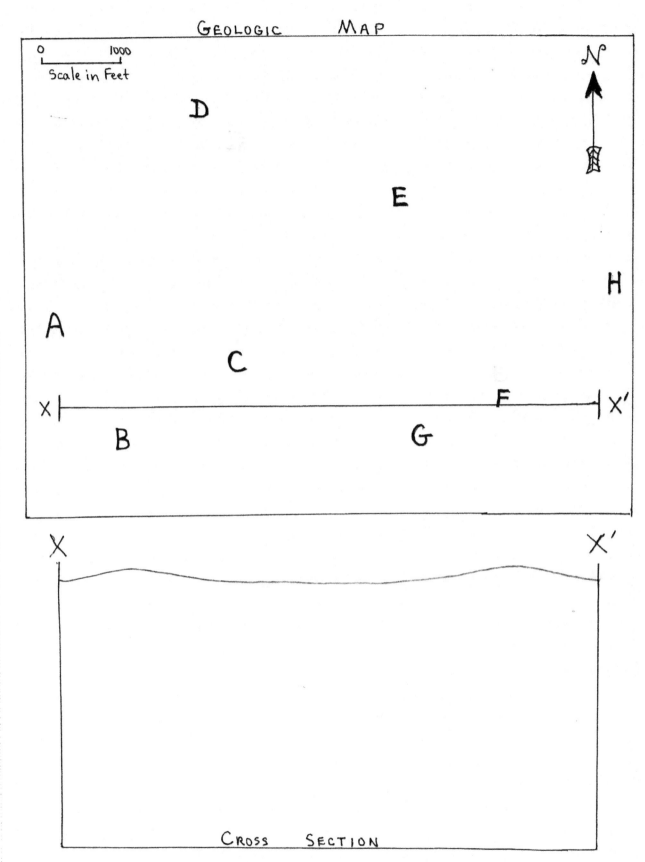

Figure 12.3. Base map for construction of Geologic Map and Geologic Cross Section (**Part 1B**).

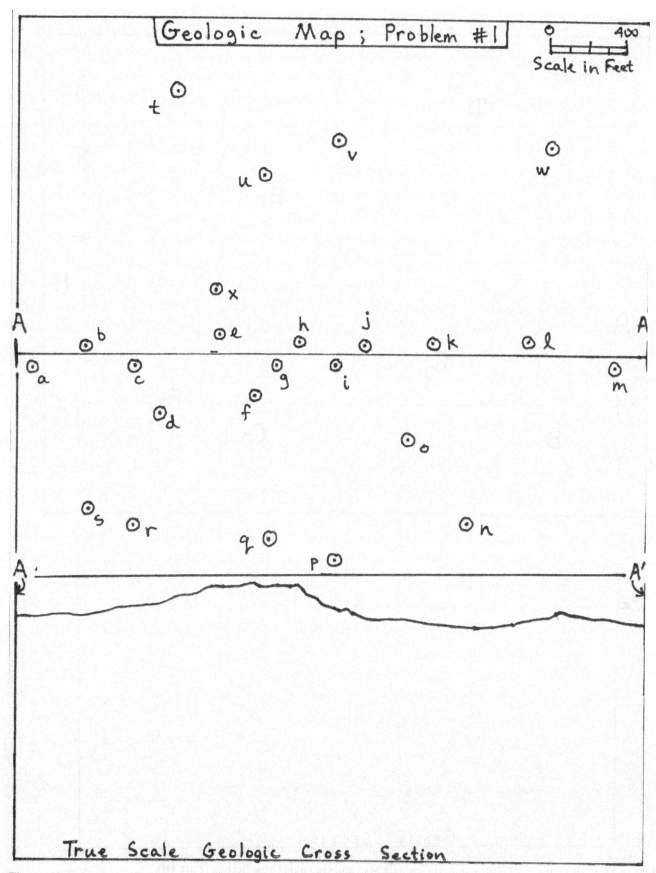

Figure 12.4. Base map for construction of Geologic Map and Geologic Cross Section (**Part 1C**).

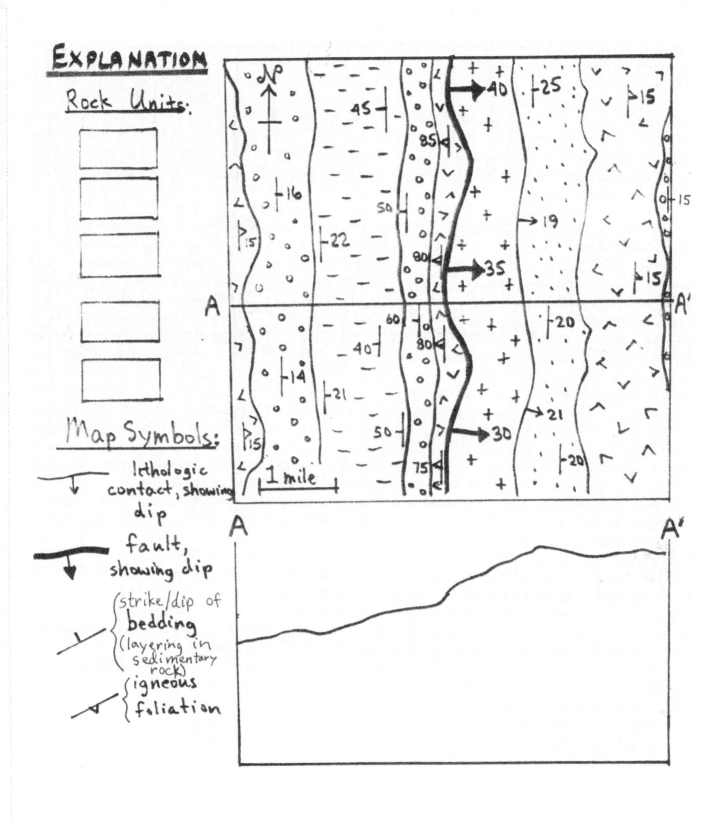

Figure 12.5. Hypothetical geologic map and cross section for Part 2. Drawing by J. Nourse.

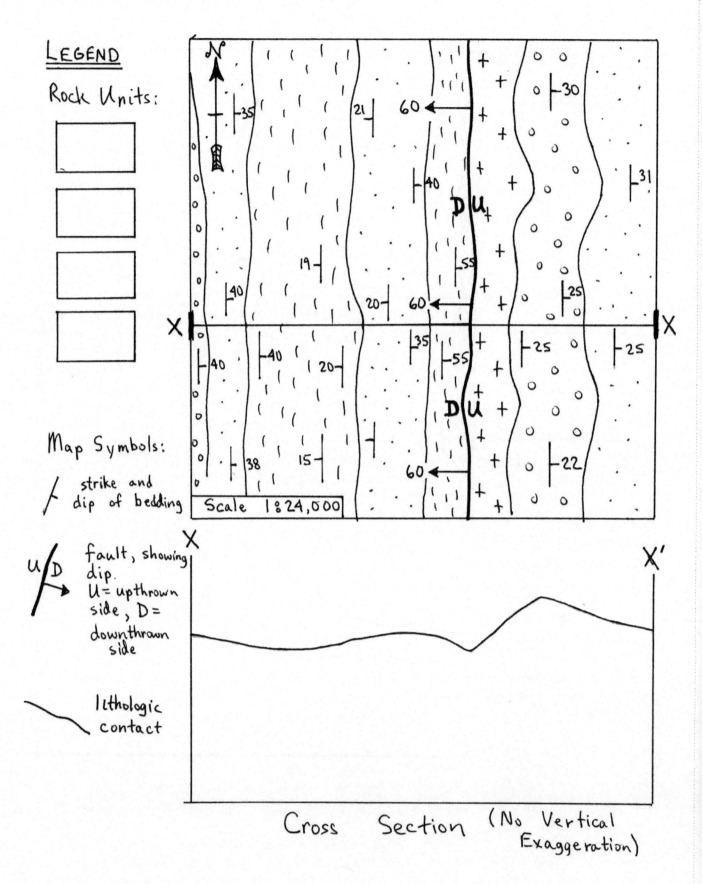

Figure 12.6. Hypothetical geologic map and cross section for Part 2. Drawing by J. Nourse.

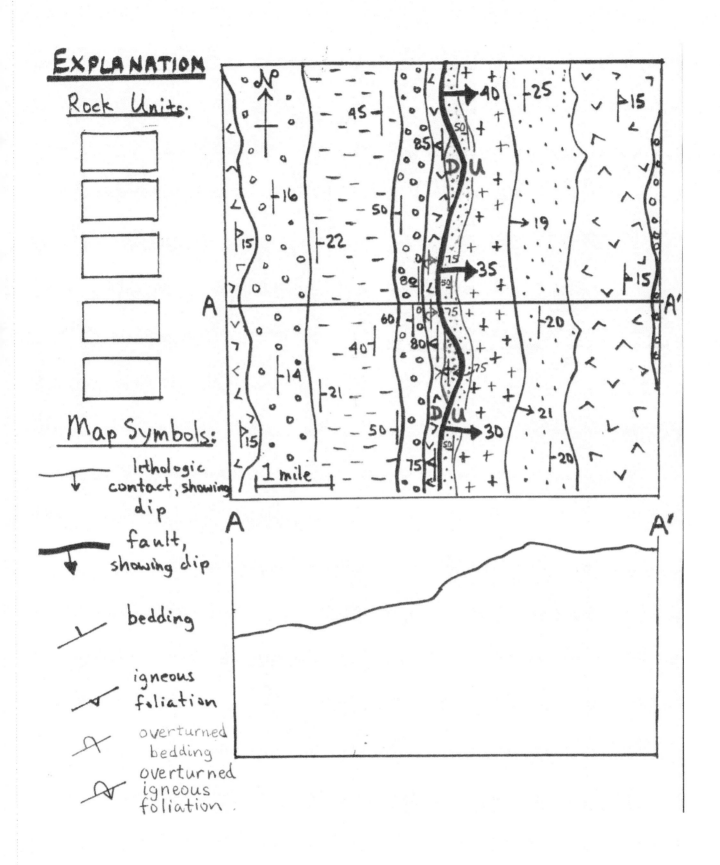

Figure 12.7. Hypothetical geologic map and cross section for Part 2. Drawing by J. Nourse.

63

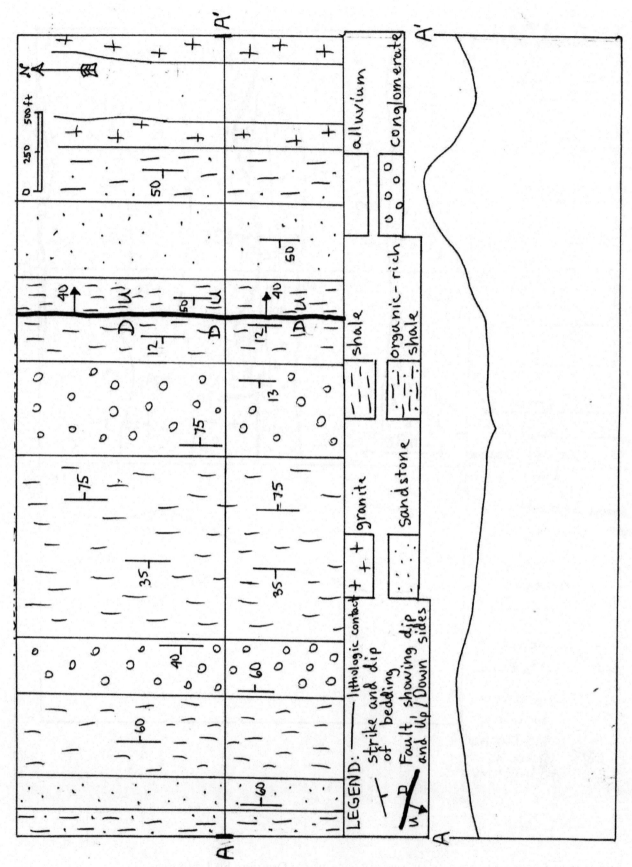

Figure 12.8.   Hypothetical geologic map and cross section for Part 2. Drawing by J. Nourse.

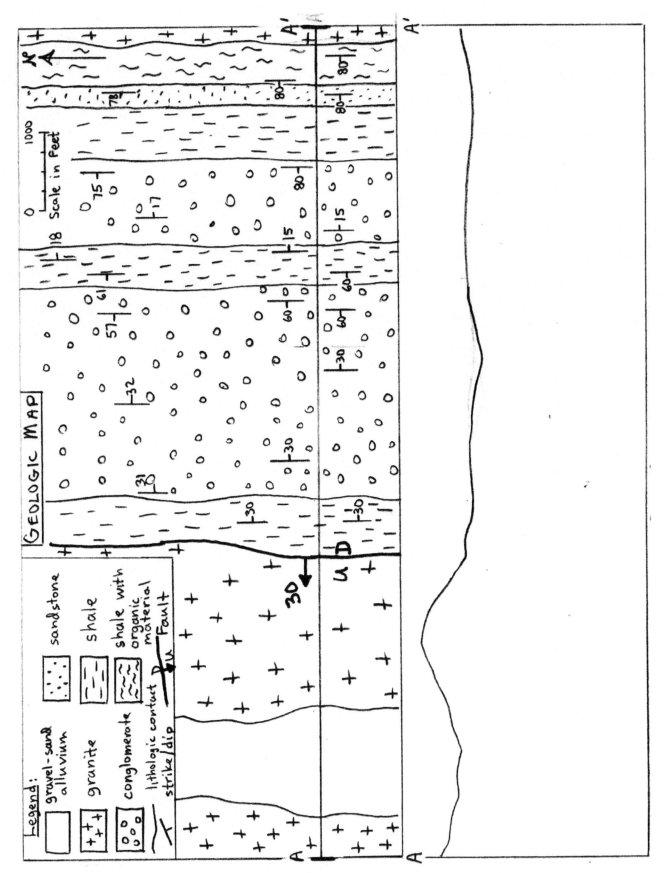

Figure 12.9   Hypothetical geologic map and cross section for Part 2. Drawing by J. Nourse.

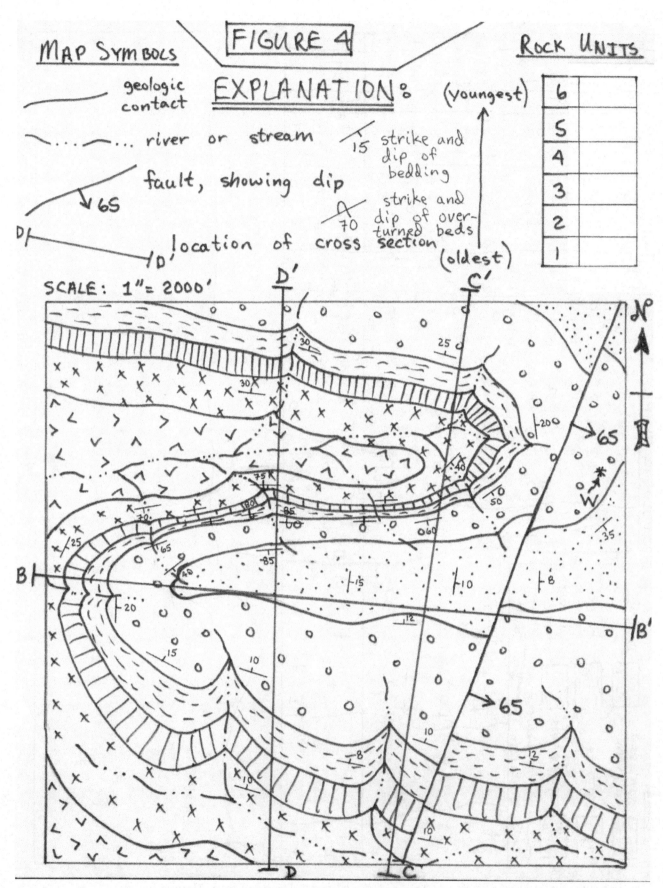

Figure 12.10a. Hypothetical geologic map for Part 2. See **Figure 12.10b** for cross sections. Drawing by J. Nourse.

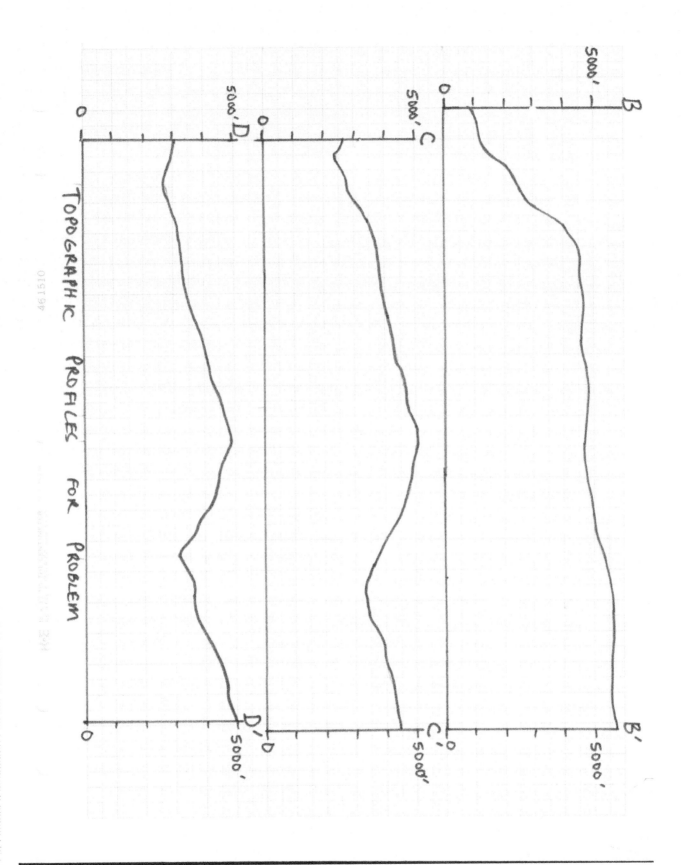

Figure 12.10b. Cross sections to accompany geologic map of **Figure 12.10a.**

Exercise #13

Name_____

Class Number_____

Geology and Groundwater Resources of Alluvial Fans

Objectives: Students will characterize alluvial fan geometry and hydrogeology by: (a) analyzing a topographic map of a typical fan system, and (b) constructing a hydro-geologic profile from topographic and borehole data. This information shall be used to calculate the volume of sediment and pore water contained in the alluvial fan.

Materials: pencil
colored pencils—blue, brown, red, green, yellow
one sheet of graph paper
ruler
calculator
working knowledge of Exercises #10 and #12

Introduction

Normal faults and reverse faults (see **Figure 18.1**) developed in arid or semiarid regions commonly separate steep mountain ranges from valleys that may be partially filled with alluvial deposits (**Figure 13.1**). These dip-slip faults usually are marked by a sharp break in slope at the mountain front. Due to an abrupt decrease of stream gradient at the mountain front, rivers exiting the mountains will deposit large volumes of gravel and sand at the edge of the valley during periodic floods. Typically this sediment forms fan shaped deposits (commonly referred to as an alluvial fans) depicted in map view by a concentric arrangement of topographic contours. Alluvial fans occur throughout the Basin and Range province of the United States. Spectacular examples may be viewed in the Death Valley area of California or generally throughout the Basin and Range Province of the American west. Throughout the world, one can expect to encounter alluvial fans in most dry regions affected by active dip-slip faulting (**Figure 13.1**) and periodic rainstorms.

Faults that bound alluvial fans may record displacements ranging from tens to thousands of meters. The long-term result is very thick deposits of porous and permeable gravel and sand that absorb large quantities of stream runoff derived from the adjacent mountains. Alluvial fans are thus prime targets for groundwater exploration. Water-saturated zones of porous ground known as aquifers provide a significant and growing proportion of domestic and irrigation water supply for residents of the American west.

Figure 13.1. Typical hydrogeologic conditions that occur in mountainous regions with semi-arid environments. Drawing by J. Nourse.

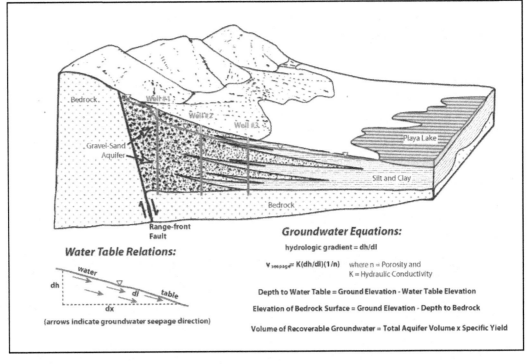

69

Procedure / Problems—please directly draw on the map or fill in the spaces below as requested

1. Study the map of the Ennis, Montana area (**Figure 13.2**) and relate to the topographic contours. Closely spaced contours in the east depict mountains underlain by sedimentary and metamorphic bedrock. More widely spaced contours in the west occupy areas covered by various types of alluvium. One prominent alluvial fan should be obvious, but careful inspection will reveal two or three more that overlap the former. At the northwest edge of the map, the Madison River flows across a relatively flat flood plain shown by widely spaced contours. Let's begin by highlighting some features and converting this image into a geologic map. Students with initiative may find it illuminating to consult http://www.mbmg.mtech.edu/pdf-open-files/mbmg543-centralmadison.pdf :

    A. Use a **blue** pencil or pen to outline the main drainage systems on **Figure 13.2**. Knowledge of the "rule of V's" for topographic contours (**Figure 10.3**) will be helpful. Notice that some streams converge as they flow from the mountains onto the fan apex, while others diverge as they flow downhill from the apex.

    B. Use a **red** pencil or pen to map in the most likely position of the normal fault system that forms a boundary between the mountains and the alluvial fans. In most cases the fault segments are marked by an abrupt change in slope as depicted by the topographic contours. Use a conventional line style for normal faults, and label the up and down sides of your fault with **U** and **D** (you may assume that the mountains have moved up, while the valley has dropped down).

    C. Use another: color (preferably **yellow** or **light brown**) to show the aerial extent of all alluvial fans (there is one large fan, but you should be able to find several others where smaller streams exit the mountains). Outline the boundaries of each fan with a **dark brown** pencil.

    D. Assuming that the mountain areas are underlain by igneous bedrock, use a contrasting color and pattern (**green pluses**, perhaps) to show the distribution of bedrock.

    E. Enhance the map legend by describing colors and symbols you have added to the map. Explain these features as necessary.

    F. Use a ruler to measure the bar scale at the bottom of the map. Express this scale as *1 inch = ???? feet*, then convert to a representative fraction. *For example*: if 1 inch on the bar scale represents a distance of 2000 ft on the ground, then 1 inch = 2000 ft x (12 inches / ft) = 24, 000 inches. Thus, 1 inch on the map = 24,000 inches on the ground so the representative fraction is 1:24,000.

2. Now let's draw a *hydro-geologic cross section* through the main alluvial fan. You may wish to refer to the procedures outlined in **Exercise #12**.

    A. Use a separate piece of graph paper (**landscape orientation**) to set up a box for constructing a topographic profile along line A-A'. Set the vertical scale of your profile to be: **1 inch = 2000 feet**   Label this vertical scale clearly.

    B. Use the contour lines to construct a topographic profile of the ground surface. Refer to **Exercise #10** for guidance.

    C. Locate the normal fault on your profile and project it to depth, assuming a dip of 75° to the west. Also show colors for bedrock and alluvium that correspond to those on your map.

3. Four wells have been drilled through the main alluvial fan. Pertinent information from well logs is summarized below. Distances corresponding to different features or earth materials represent *depths* below the ground surface:

    Well #1

Total depth drilled (*depth = distance measured down from ground surface*)	2880 ft
Elevation of ground surface	5980 ft
Coarse gravel with some coarse sand layers; pores not saturated	0-100 ft
Water table	100 ft
Water saturated coarse gravel with some coarse sand layers	101-2780 ft
Non-porous igneous bedrock	2780-2880 ft

Well #2

Total depth drilled	2100 ft
Elevation of ground surface	5600 ft
Fine gravel with coarse sand layers; pores not saturated	0-200 ft
Water table	200 ft
Water saturated fine gravel with coarse sand layers	200-2000 ft
Non-porous igneous bedrock	2000-2100 ft

Well #3

Total depth drilled	1580 ft
Elevation of ground surface	5280 ft
Medium sand with fine sand and some silt; pores not saturated	0-150 ft
Water table	150 ft
Water saturated medium sand interlayered with fine sand and some silt	150-1480 ft
Non-porous igneous bedrock	1480-1580 ft

Well #4

Total depth drilled	1050 ft
Elevation of ground surface	4950 ft
Silt and clay with some fine sand layers; pores not saturated	0-50 ft
Water table	50 ft
Water saturated silt and clay interlayered with fine sand	51-950 ft
Non-porous igneous bedrock	950-1050 ft

A. Locate the four wells on your topographic profile of Problem #2. For each well, draw a vertical line extending from the ground surface to the elevation where the well bottoms out in bedrock.

B. Plot the position of the water table in each well. Connect these four points with a continuous dark blue line. Draw a series of inverted triangles on the water table to distinguish it further.

C. Use patterns suggested by your instructor to indicate the type of sediment or rock intersected by each well. The main idea is to show differences in particle size. Then try to fill in the space between the four wells by extrapolating the patterns. Generally, you can expect the different soil units to "interfinger" such that the grain sizes get smaller as you move westward from the mountains to the flat part of the valley (**see Figure 13.1**). Add a legend to your cross section.

D. What is the vertical exaggeration of your completed hydro-geologic cross section?
Hint: Vertical Exaggeration = Horizontal Scale / Vertical Scale

4. At this stage you should have developed a fairly coherent three-dimensional representation of the main alluvial fan. It is clear that a fairly thick aquifer exists beneath the surface of the fan. An aquifer is defined to be any body of porous, water-saturated ground that has sufficient permeability to transmit water to a well. Let's try to quantify the volume of the alluvial fan and the amount of groundwater contained in its pores. **Please perform your calculations below your cross section:**

A. Use your knowledge of geometry, dimensions from your map and hydro-geologic profile, and the values of the topographic contours to calculate the total volume of sediment that comprises the main alluvial fan east of the 5200 foot contour. Your instructor will provide some clues about how to tackle this problem. Considering the fairly regular shapes involved, you should be able to makes some time-saving approximations instead of applying the method of horizontal slices. Show all of your work below, using additional pages as necessary. Express your answer in **cubic feet**.

B. Assume that the average porosity of the fan is 30%. Also assume that the water table geometry illustrated in your hydro-geologic profile is similar for other downhill slices through the fan. Now calculate the total volume of groundwater contained in the pores of the main alluvial fan. Show your work below, using additional pages as necessary. Express your answer in **cubic feet**. *Hint*: Recall that **Porosity = Volume$_{voids}$ / Voume$_{bulk}$** Determine the bulk volume of water-saturated sediment present below the water table. Then, solve the above equation for Volume$_{voids}$. This is equivalent to the volume of pore water.

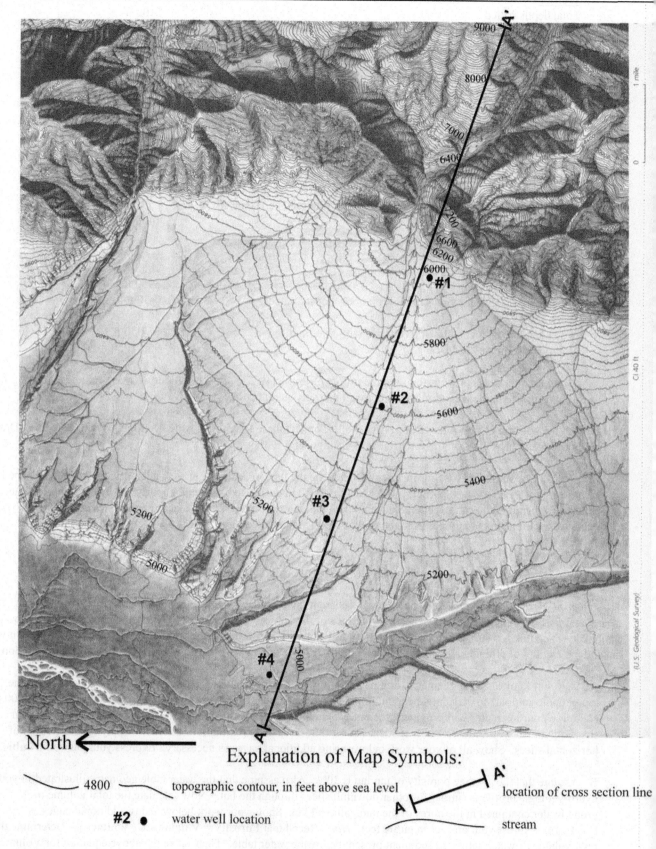

North ◄─────────────

Explanation of Map Symbols:

～～ 4800 ～～ topographic contour, in feet above sea level

#2 ● water well location

A├───────┤A' location of cross section line
A'

～～～～ stream

Figure 13.2 Topographic map of alluvial fans near Ennis, Montana. Modified from Figure 17, p. 110, of Hamblin and and Howard (1999).

Exercise #14

Name_____

Class Number_____

Water Table Maps, Groundwater Flow, and Seepage Velocity in Unconfined Aquifers

Objectives   To understand the dynamics of groundwater flow in unconfined aquifers. Students will construct water table contour maps and use this information to determine depth to the water table, the direction of groundwater flow, and the velocity of groundwater seepage.

Materials   pencil
ruler
colored pencils (blue, red, green, brown)
calculator
graph paper

Introduction

Unconfined aquifers, also known as "water table aquifers," provide the majority of the world's fresh groundwater resources. In general, an *unconfined aquifer* is any water-saturated zone of porous and permeable ground that is bounded on top by a water table, and on the bottom by substantially less permeable ground. As illustrated below in **Figure 14.1**, the location of the water table coincides with the elevation of standing water in a lake, stream, or shallow water well. The water table is a three-dimensional surface that can be contoured using the same procedures as those outlined in **Exercise #10**. Once contours have been drawn, it is a simple matter to determine the direction of groundwater seepage. If one knows the porosity (**n**) and permeability (hydraulic conductivity; **K**) of the aquifer, the seepage velocity ($v_{seepage}$) may be calculated. Finally, if the aquifer dimensions are known, the total groundwater discharge (**Q**) can be determined using a famous equation known as *Darcy's Law*. Students will have an opportunity to work with these relationships, summarized below:

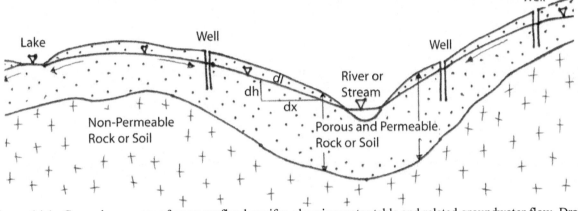

Figure 14.1. General geometry of an unconfined aquifer, showing water table and related groundwater flow. Drawing by J. Nourse.

General Rules Regarding Water Tables and Groundwater Flow:

(1) In most situations, water table elevation is equivalent to:
(a) the surface elevation of lakes
(b) the surface elevation of any point on a flowing stream
(c) the elevation of standing water in a shallow well

(2) Groundwater will flow in the direction of the steepest hydraulic gradient
(a) from high elevations toward low elevations
(b) perpendicular to water table contours that define the gradient

(3) The volume and velocity of groundwater seepage will be directly proportional to the hydraulic gradient and the permeability

73

General Equations Pertaining to Groundwater Flow:

Depth to Water Table = Ground Elevation – Water Table Elevation

Porosity = n = Volume$_{voids}$/Volume$_{bulk}$

Hydraulic Gradient = dh/dl = dh/dx (for cases where slope of water table is small)

Seepage Velocity = $v_{seepage}$ = $(1/n)K\ (dh/dl)$

Groundwater Discharge = Q = $KA(dh/dl)$

A = water-saturated Area oriented perpendicular to groundwater flow

Procedure / Problems   Please draw requested features directly on the maps. Show all calculations and other work on separate sheets of paper. Add a legend to each map (or enhance the existing legend), explaining pertinent symbols:

1. Refer to **Figure 14.2** Your instructor will guide you through this Practice Exercise to illustrate procedures for contouring water tables and constructing hydrologic profiles.
    A. Use a blue pencil to highlight all surface locations of known water table elevation. These will be all the lake surfaces and specific points along the perennial stream.
    B. Make a water table map with a contour interval of 10 feet, beginning with the 90 ft contour. The procedure is similar to that of **Exercise 10, Part 2**. Start with regular pencil, then use a **dark green** pencil or pen to color each contour line.
    C. Construct sufficient flow lines to illustrate the groundwater flow field on your map. These should be red lines with arrows; each constructed to intersect the water table contours at 90 degrees. Arrows point in the down-gradient direction.

2. Refer to **Practice Exercises #4A** and **#4B** reproduced in **Figure 14.3**. This activity illustrates the concept that water table contours "V" upstream for in the vicinity of *gaining streams* and "V" downstream in the vicinity of *losing streams*.
    A. Use the "rule of V's" (**Exercise #10**) to locate the positions of flowing streams on both drawings.
    B. Contour the water table elevation data in #4A and #4B. **Refer to Problem 14.1**; Use a contour interval of 10 ft.
    C. Construct sufficient flow lines to illustrate the groundwater flow field on each map.
    D. Classify the streams as "gaining" or "losing." Do they gain water from or lose water to the ground?
    E. Determine the depth to the water table at points A-E.   (**Depth = Ground Elevation – Elevation of Water Table**)
    F. Construct two *hydrologic profiles* (cross sections showing both the land surface and the water table) along lines A-A' and B-B'. Use a sharp pencil and an exaggerated vertical scale of 1 inch = 50 ft. Start with the topography (land surface), marking elevations for the contours, stream crossings, and cross section end points. Assume that the lakes in Exercise #4A have steep shorelines (cliffs) that extend 5 ft above the lake surface. Then mark in the water table, using stream elevations, water table contours, and lake elevations for control. Use a blue pencil or pen to indicate the water table on each profile and show groundwater flow direction with arrows. Interpolate the water table elevations at the edge of each cross section

3. Refer to the hydrologic map in **Figure 14.4**.
    A. What is the elevation of the lake (to the nearest 5 feet)?
    B. Contour the water table data, using a contour interval of 10 ft.
    C. Construct sufficient flow lines to illustrate the groundwater flow field on the map.
    D. What is the depth to water at point X?
    E. Use the space below the map to construct a hydrologic profile along an east-west line through center of the lake.

4. Refer to the hydrologic map in **Figure 14.5**.
    A. Contour the water table data, using a contour interval of 10 ft. **Refer to Problem 14.1 for guidance.**
    B. Construct flow lines emanating from each point "x" to illustrate the groundwater flow field on the map.
    C. Will the stream flow at gauge A be <u>greater</u> than or <u>less</u> than at gauge B?
    D. What is the depth to the water table at Points Y and Z? (**Depth = Ground Elevation – Elevation of Water Table**)
    E. What is the hydraulic gradient (dh/dl) at Point Y?
    F. Suppose you know that the aquifer has porosity, n, of 33% and hydraulic conductivity, K, of 10 ft/day. How long will it take water to seep from Point Y to the edge of the map? *Hint*: one of the equations noted above will be helpful.

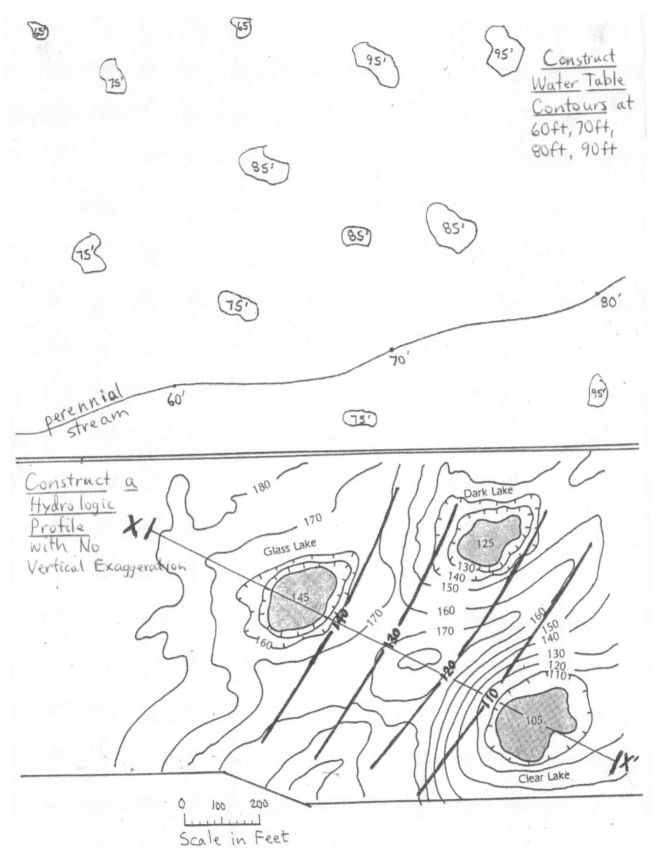

Figure 14.2. Practice exercise for learning concepts of water table contouring and hydrologic profile construction. The lower figure is modified from Figure 11.2, page 118 of Hamblin and Howard (1999).

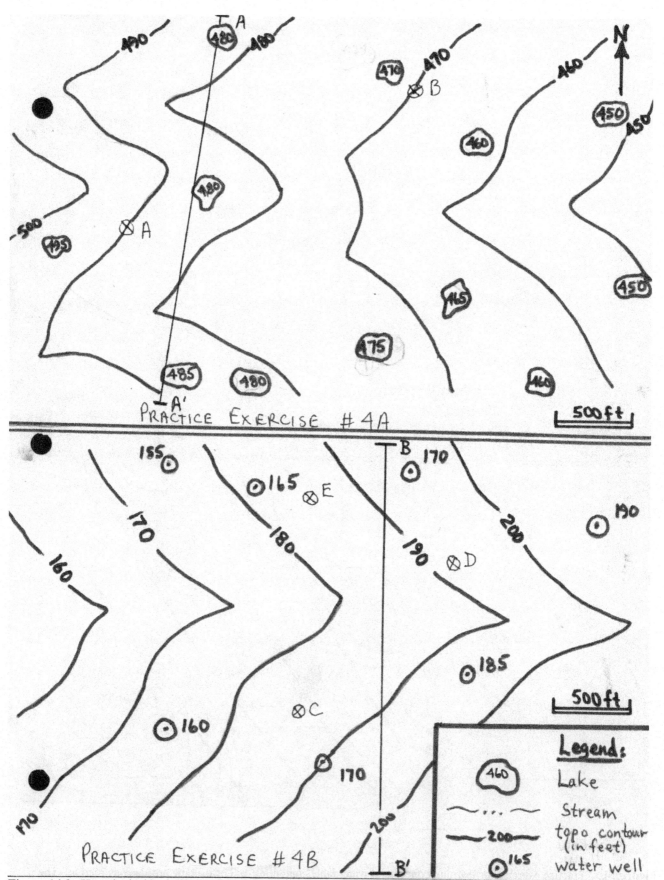

PRACTICE EXERCISE #4A

PRACTICE EXERCISE #4B

Figure 14.3. Hypothetical hydrologic relations near two flowing streams. Drawing by J. Nourse.

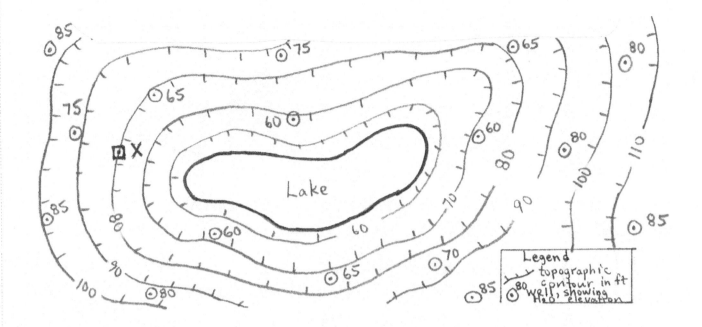

Figure 14.4. Hypothetical hydrologic relations near a lake in a wet region, where the lake level is sustained by groundwater seepage from the surrounding ground. Drawing by J. Nourse.

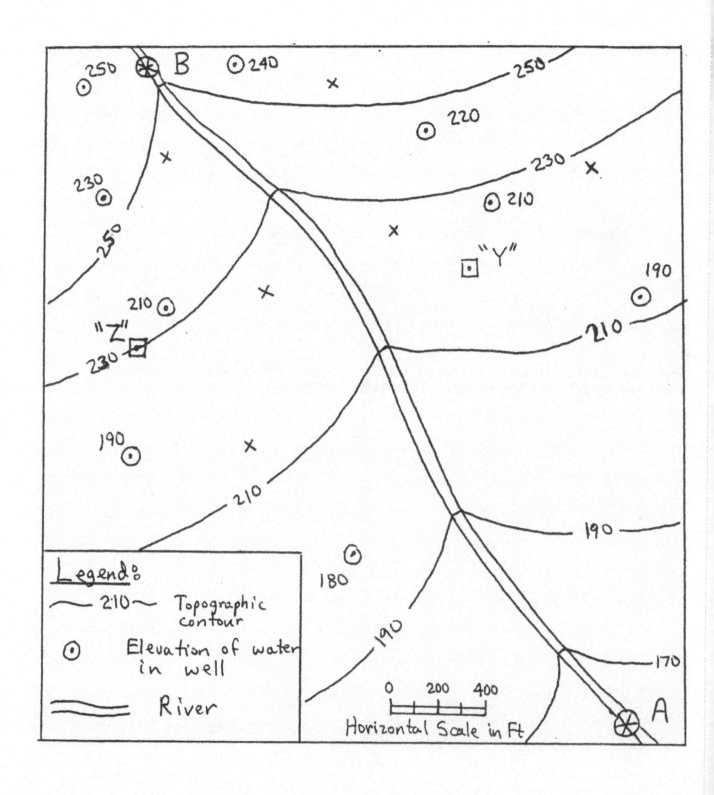

Figure 14.5. Hypothetical hydrologic relations near a flowing river during a period of extended drought. Points **A** and **B** represent flow gauges on the river. Drawing by J. Nourse.

Exercise #15

Name_____

Class Number_____

Drawdown and Subsidence When Pumping From Unconfined Aquifers

Objectives Students will apply the principle of superposition to determine water table drawdown resulting from the combined effects of pumping from three wells. Then the resultant drawdown curve shall be utilized to calculate subsidence at various points along a vertical profile.

Materials pencil
ruler
compass (for drawing circles)
colored pencils (blue, red, green, brown)
calculator

Introduction

Unconfined aquifers provide a significant source of water for drinking and irrigation in many areas. This kind of aquifer is extensively developed because water (if present) usually can be found close to the surface (within a couple hundred feet or so). An *unconfined aquifer* is defined as an underground body of porous and permeable, water-saturated material bounded on the bottom by less permeable rock and bounded on the top by a water table. The water table is a surface that separates ground with water-saturated pores (below) from ground with unsaturated pores (above). Theoretically, when drilling a well into an unconfined aquifer, water will first be encountered at the depth of the water table. If the aquifer has sufficient permeability and porosity, the well can be pumped to yield a usable quantity of water. Here is where the problem begins.

Pumping water from a well may cause the water table to be lowered. This effect is referred to as *drawdown.* The geometry of water table drawdown is illustrated in **Figure 15.1**. Pumping from a well creates a conical shaped depression in the water table, centered about the well. This *drawdown surface*, above which the water has been removed from the pores, is known as a *cone of depression*. The cone of depression grows with time, but under ideal conditions, will stabilize. This *steady-state* condition occurs when the rate of groundwater recharge (Q_{gw}) from the sides of the well balances the rate of withdrawal from the well (Q_{pump}). The three-dimensional shape of the water table during drawdown is a complex logarithmic function of time (**t**) and radial distances from the well; it also depends on the permeability and storage capabilities of the aquifer. Fortunately, the shapes of the drawdown curves used in this exercise have already been calculated. Because this problem will assume steady state conditions, the only numbers you will need to keep track of are the values of drawdown (**s**) at different radial distances.

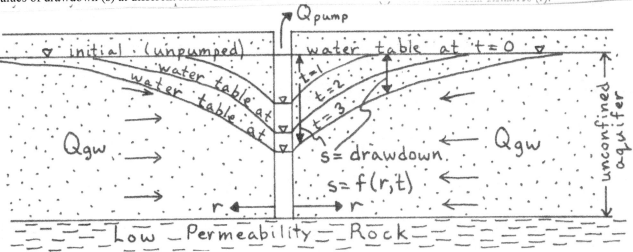

Figure 15.1 Cross section of a typical unconfined aquifer illustrating the time-dependent geometry of water table drawdown associated with pumping from a well. Drawing by J. Nourse.

Drawdown effects can become a real problem when more than one user is tapped into the same aquifer. Each well that pumps from the aquifer will produce a separate cone of depression. Where these cones overlap the drawdown effects are

<u>additive</u> and the resultant water table will be lower than the drawdown induced by either of the wells individually. Hence, excessive pumping by your neighbor could cause your own well to go dry. Imagine the legal battles that are fought over groundwater rights in areas where aquifers have been over-pumped!

One can predict the drawdown effects of pumping from multiple wells by applying the *principle of superposition.* This principle uses the concept that separate cones of depression can be added together to produce a *resultant,* or *composite drawdown curve.* The procedure for superimposing two wells is outlined below in **Figure 15.2.** Simply add the drawdown values at a given position to determine total drawdown (**s**), then show this magnitude of drawdown on the profile. Use the vertical scale to help you position several points along the section where the water table is lowered.

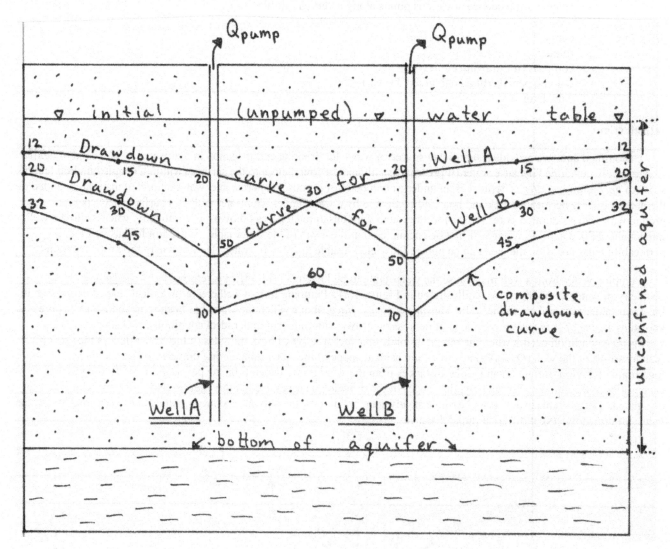

Figure 15.2. Method of superposition used to construct the **composite drawdown curve** resulting from two wells (**A** and **B**) that pump from an unconfined aquifer at the same known rate (**Q**pump). Drawing by J. Nourse.

Questions / Problems

Figure 15.3 shows the positions of three wells (**Well A, Well B,** and **Well C**) on a map of an unconfined aquifer. Each of these wells has been pumped at a different rate for a long time such that steady-state conditions currently exist. Drawdown information for the three wells is given below:

Radial Distance from Well	0m	200m	400m	600m	800m	1000m	1200m
Drawdown at Well A	100m	50m	32m	18m	8m	4m	2m
Drawdown at Well B	150m	75m	48m	27m	12m	6m	3m
Drawdown at Well C	50m	25m	16m	9m	4m	2m	0m

1. Study the horizontal scale on the map (**Figure 15.3**), then use your compass to construct drawdown contours, centered about **Well C**, for various radial distances (**r = 0 m to r = 1000 m**). These contours will form a series of concentric circles on the map. Label each contour with an appropriate drawdown value in meters.

2. Construct a cone of depression showing the drawdown (**s**) associated with **Well C** pumping alone. Do this by projecting the drawdown from each of your <u>map</u> contours to the appropriate elevation on the <u>cross section.</u> Note that the drawdown is subtracted from the initial (non-pumping) water table elevation to determine the elevation of the drawdown curve at each position. Use red to outline the cone of depression associated with Well C.

3. Now use your compass to construct drawdown contours for **Wells A** and **B** on the map. Label each contour with an appropriate drawdown value.

4. Add together drawdown values for all three wells at eight or ten different positions **where they are intersected by the line of cross section.** Write these <u>*total drawdown*</u> values on the map.

5. Project the **total drawdown** values from the map to the cross section. Reference to **Figure 15.2** may be helpful here. Use a blue curve to connect the dots. Label each dot with a value of total subsidence. You have now constructed a *composite drawdown curve* that shows the combined effects of pumping from **Wells A, B**, and **C** <u>along this particular line of cross section.</u>

6. Is **Well C** deep enough to accommodate the pumping effects of Wells A and B? Explain.

7. Suppose you have a choice of living at **Site X** or **Site Y**. Use the principle of superposition along with the drawdown contours on the map to determine the <u>elevation</u> of the water table under each of these sites. Show your work in the space below. Theoretically, you would have to drill to this level before hitting groundwater.

8. Rapid withdrawal of pore water from fine-grained sediments is known to induce <u>*subsidence*</u>, or sinking of the ground. In general, subsidence will be proportional to total drawdown of the water table at a given location. Let's assume a subsidence value of 10% for this region. In other words, the ground surface would drop 10 m for each 100 m of total drawdown. Go back to Problem #5 above, and calculate a subsidence value corresponding to each value of total drawdown. Then plot one more curve representing subsidence. Simply measure the vertical distance **downward from the ground surface** at each point where you know the subsidence. Then connect the dots, using a sharp red pencil or pen. Label each dot with a specific value of subsidence.

9. Make a simple **legend** to accompany **Figure 15.3**. Explain all new symbols (preferably color-coded) that you have added.

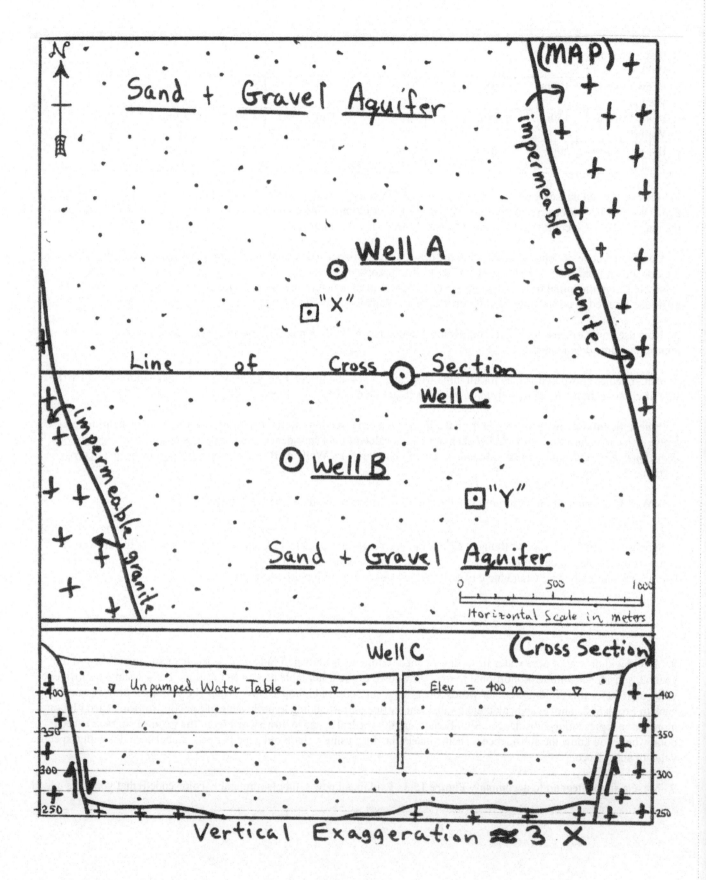

Figure 15.3. Geologic map and cross section of an unconfined aquifer subjected to steady-state pumping from three wells. Drawing by. J. Nourse.

Exercise #16

Name: _____

Class Number: _____

Volcanic Hazards

Objectives: To learn about the volcanic hazards associated with a particular active volcano. Students will perform an Internet search to determine the eruption history and the distribution of volcanic hazards as mapped by volcanologists.

Materials: pencil and paper
colored pencils or pens
access to the Internet and a color printer

Introduction

     The photographs on the next page (**Figures 16.1-16.5**) illustrate the awe-inspiring power and beauty of some famous volcanoes. Implications for people who live in the vicinity should be obvious. In this exercise, you will explore some excellent web sites that describe human impacts associated with an active volcano of your choice.

Procedure / Problems

     Choose one of the following active volcanoes for closer analysis. Your instructor may have additional suggestions, or, you could pick your own favorite volcano:

     Mt. Saint Helens; Mt. Rainier; Mt. Lassen; Mt. Baker; Mt. Pinatubo (Philippines); Mt. Vesuvius (Italy); Galeras Volcano (Columbia); Popocatepetl (Mexico); various Hawaiian volcanoes; various Iceland volcanoes

Check out the USGS volcano websites: Volcano Hazards Program https://volcanoes.usgs.gov/index.html ; Cascade Volcano Observatory http://vulcan.wr.usgs.gov; Hawaiian Volcano Observatory http://hvo.wr.usgs.gov Explore the many links. Also perform a general Internet search using Google or some other search engine. Then complete the following problems:

1. Outline the eruption history of the volcano you have chosen. Include noteworthy events that have had <u>major</u> human impact.

2. Give the name and population of the major city (or cities) that could be adversely affected by an eruption of your chosen volcano.

3. Print out (in color) a _volcanic hazards map_ published by USGS volcanologists or other reputable scientists. Many are available for downloading as PDF or GIF files. Be sure to cite the author and date of the map. Add some color to highlight pertinent features as needed.

4. List the primary hazards associated with your volcano, then briefly describe their causes and lateral extent. It may be helpful to read portions of the associated hazards report.

5. Print out your favorite picture of the volcano and attach it to the front of your report. Again, be sure to cite the source of the photograph.

Figure 16.1. 1980 eruption of Mt. Saint Helens
Photograph from the USGS Cascades Volcano Observatory.

Figure 16.2. Mt. Saint Helens crater, viewed from the north.
Photograph from the USGS Cascades Volcano Observatory.

Figure 16.3. 1991 eruption of Mt. Pinatubo, Philippines.
Photograph by J. N. Marso, USGS, July, 1991.

Figure 16.4. Eruption of Vatnajokull, Iceland volcano in 1996, through glacial ice. This event caused severe flooding in the region. Photograph by Magnus Tumi Gudmondsson.

Figure 16.5. Popocatepetl awakens in the mid 1990's
Photograph by John W. Ewert, USGS.

Exercise #17

Name: _____

Class Number: _____

Rivers and Flood Hazards

Objectives: To develop an understanding of the natural flood behavior of rivers and to learn how to evaluate flood hazards

Materials: ruler
colored pencils: red, green, blue; black ink pen
magnifying lens (to read contours on Figure 17.1)
calculator

Introduction

In late June of 1972, Hurricane Agnes swept across the northeastern United States producing one of the worst floods in U.S. history. This flood killed 75 people and caused nearly $2.8 billion in damage. The most significant devastation occurred within the watershed of the Susquehanna River (**Figure 17.6**), which flows from its headwaters in southern New York and northern Pennsylvania to the Chesapeake Bay in Maryland. During the flood's peak, river flows (measured as both _stage_ and _discharge_) were higher than any other Susquehanna River flood since record-keeping began in 1892 (Merritts et al., 1998).

In the wake of such a devastating flood, residents of the Susquehanna watershed demanded improved flood protection. In order to reduce the hazards of future flooding, scientists, engineers, and urban planners needed to answer several important questions (see also Merritts et al., 1998):

What was the maximum height of the flood water during this event?

How often do large floods like this occur on the Susquehanna River? (this is referred to as the _recurrence interval_)

How did the timing and magnitude of the flood peak vary at different locations along the Susquehanna River drainage network?

To answer these questions, you will examine a topographic map of one flooded area (the city of Harrisburg, Pennsylvania) and you will use discharge data from the Hurricane Agnes flood to plot flood hydrographs and flood hazard zones. Please write your answers on additional sheets of paper as necessary.

Procedure / Problems

Part A. Mapping Flood Hazards at Harrisburg

Using the topographic map of Harrisburg (**Figure 17.1**), answer the following questions:

1.   The Harrisburg gauging station is located on the east bank of the Susquehanna River near the southeastern corner of the map. The gauge reads 0' when the water elevation, or _stage_, is 290' above sea level. During the peak of the Agnes flood in 1972, the stage height at the gauging station was about **32.5 feet**. What was the peak water surface elevation above sea level during the 1972 flood?

2.   On the topographic map, trace with a **red pencil or pen** the approximate limit of the Hurricane Agnes flood waters on both sides of the river. To do this, you will draw a contour line that corresponds with the water surface elevation that you determined in question 1. _Hint_: This has been done for you already; just highlight the line in red. Use a **blue pencil** to **lightly** color in the area that was covered by water in 1972.

3.   Examine the floodplain through downtown Harrisburg on the east side of the Susquehanna River. Also check out the islands. What important buildings were flooded in 1972?

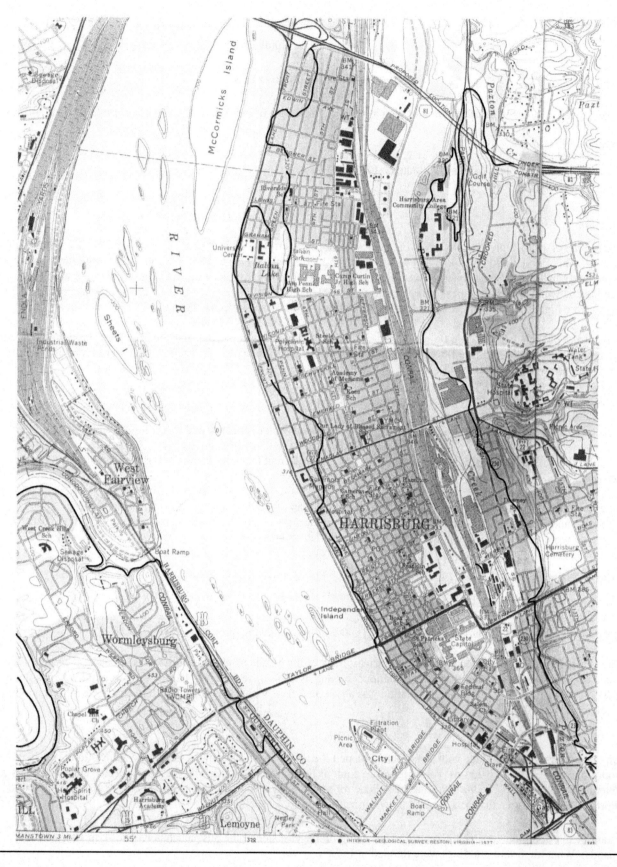

Figure 17.1. Topographic map of the Susquehanna River at Harrisburg, Pennsylvania. The 320 ft contour is highlighted; this corresponds approximately to the maximum extent of the 1972 flood. The shoreline of the river shown on this map is at 290 ft elevation.

86

4. Do you think it is a good idea to build important public facilities on a floodplain? Why or why not? What types of facilities might be difficult to relocate (those that require a riverside location)?

5. Prior to Hurricane Agnes, the largest flood on record at Harrisburg was that of 1936. During the flood's peak, the stage height at the gauging station was about 28.5 feet. What was the water surface elevation above sea level during the 1936 flood?

Part B. Flood Frequency and Recurrence Intervals

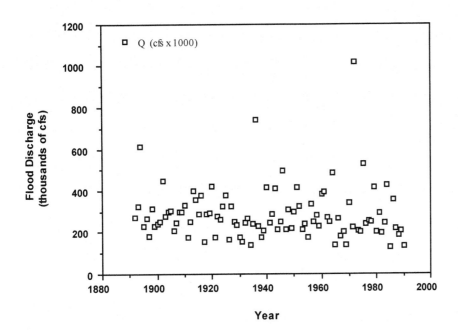

Figure 17.2. Graph showing the discharge of maximum annual floods at Harrisburg, PA from 1892 to 1991.

1. Examine the graph in **Figure 17.2**. What has happened to the magnitude of the largest floods (>600,000 cfs) over the last century? What about the magnitude of the medium-sized floods (300,000 - 600,000 cfs)?

2. What are some possible reasons for the observed changes in flood magnitudes?

The flood history of a river reveals important clues about the size and frequency of floods that can be expected in the future. Longer historical records provide more accurate statistical data about the river's long-term flood behavior and its probable future activity. The *Recurrence Interval (RI)* for a particular size flood represents a probability of that flood occurring within a specific time period. The graph in **Figure 17.3** shows a Flood Frequency curve for the Susquehanna River at Harrisburg. Recurrence intervals were calculated using the "Weibull formula":

$$RI = \frac{n + 1}{m}$$ where n = total number of years in the historical record (100 years in this case)
m = rank of a flood discharge in a list of maximum annual floods

87

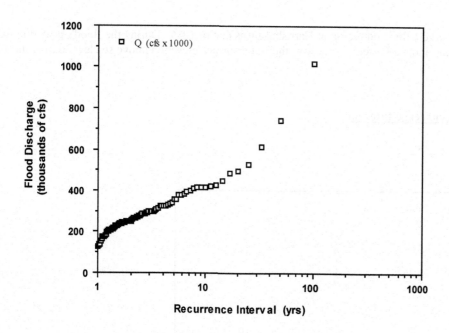

Figure 17.3. Flood frequency graph for the Susquehanna River at Harrisburg, PA.

3. Examine **Figure 17.3**. What general trend do the data points follow? (Ignore the three largest outlying floods in the upper right hand corner of the graph)

4. Draw a best-fit line through the data. (Again, ignore the three largest outlying floods)

5. Figure 16.3 shows the Hurricane Agnes flood as a 100-year flood. Is this correct? What is the actual recurrence interval for a Hurricane Agnes size flood? Show on the graph how you determined this value. What is the true recurrence interval for the 1936 flood? Show your calculation below:

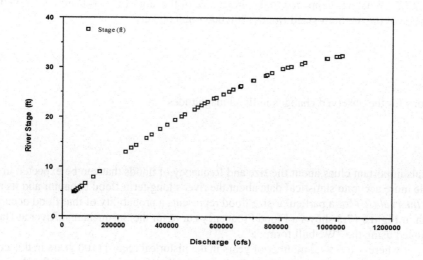

Figure 17.4. Rating Curve for the Susquehanna River at Harrisburg, PA.

4. Figure 17.4 shows a *Rating Curve* for the Harrisburg gauging station. If you were in charge of building a flood wall along the river to protect downtown Harrisburg from the **10 year flood**, how high would it have to be? How about for the **50 year flood**?

C. Flood Hydrographs for Hurricane Agnes

A hydrograph is a plot of river stage (elevation above gauging station datum) versus time. River discharge (volume/time) can also be substituted for stage, as the two are proportional to one another. **Figure 17.5** shows Hurricane Agnes flood hydrographs for three different locations in the Susquehanna River Watershed. The three sites, moving progressively downstream, are: (1) Clearfield Creek at Dimeling, PA,(2) the West Branch Susquehanna River at Williamsport, PA, and (3) the main trunk of the Susquehanna River at Harrisburg (see watershed map of **Figure 17.6**). These three sites each depict a different flood behavior as the flood wave moves downstream over time.

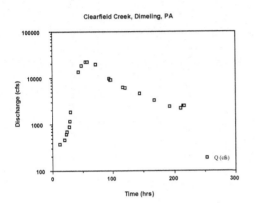

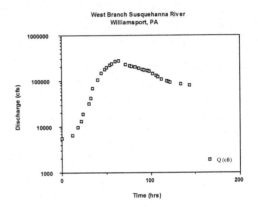

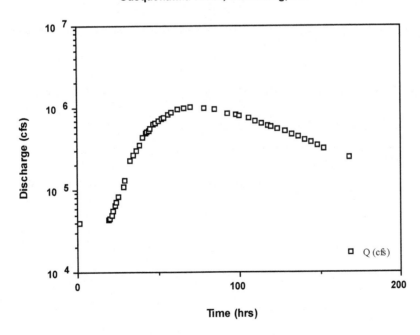

Figure 17.5. Hurricane Agnes flood hydrographs for three locations in the Susquehanna River watershed.

89

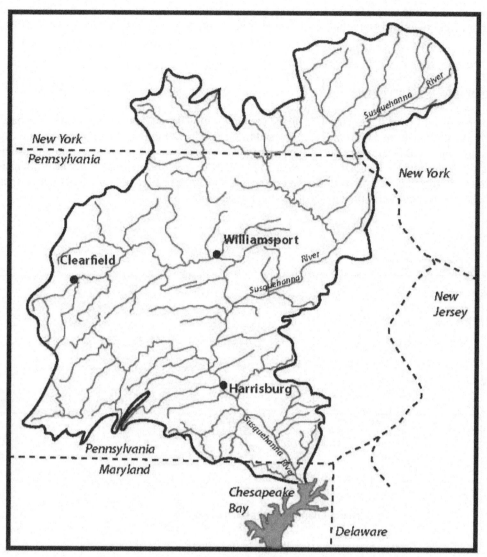

Figure 17.6. Map of the Susquehanna River watershed

1. Please use a **blue pencil or pen** on the map above to outline the Susquehanna River and its main tributaries. Also use a **red pencil** to highlight locations of the Clearfield Creek, Williamsport, and Harrisburg gauging stations.

2. Using the data from **Figure 17.5**, fill in the data table below:

Gauging Station	Drainage Basin Area	Peak Discharge	Time of Peak Discharge
Clearfield Creek (CC)	371 sq. mi.		
Williamsport (Wm)	5,682 sq. mi.		
Harrisburg (Ha)	24,100 sq. mi.		

3. What is the relationship between the magnitude of peak discharge, the station location within the watershed, and the drainage area?

4. What is the relationship between time of peak discharge, station location, and drainage area?

90

Fault Recognition and Classification;
Seismic Hazards Zoning; Building in Earthquake Country

Objectives: (a) To recognize the map and cross-section expression of typical dip-slip and strike-slip faults.

                 (b) To utilize crosscutting age relations to assess whether a fault is active or inactive.

                 (c) To gain experience with map zoning of seismic hazards.

                 (d) To learn relationships between foundation type and intensity of seismic shaking

Materials: pencil

                 ruler

                 colored pencils: red, blue, green, or brown

                 black ink pen

                 one piece of tracing paper

                 working knowledge of "laws of crosscutting relationships" (Exercise #7)

Introduction

Active faults are commonly manifested as *scarps* or *topographic anomalies* visible at earth's surface. These anomalies may be represented by distinct contour patterns or drainage networks on topographic maps. Geologists who specialize in *geomorphology* and *neotectonics* are trained to recognize fault-related landscape features as well as landslides, volcanic vents, and other imprints of geocatastrophic events. Accurate maps and aerial photographs provide them a most valuable data resource. The first parts of this exercise will address map patterns and geologic offsets that typically result from three types of faults (**Figure 18.1**). *Dip-slip faults,* which are broken into *normal* and *reverse* geometries, cause one side of the fault to move up relative to the other. These faults therefore tend to form sharp discontinuities between mountains and valleys. *Strike-slip faults* cause one side of the fault to slide horizontally (sideways) past the other. In this case, mountains are not directly associated with the fault movement. Instead, linear zones of crushed rock near the fault tend to be more easily eroded, and river drainages may be incised along the same trends. This concept is addressed in **Problem #1**.

One important aspect of seismic hazards assessment is determining when a fault was last active. This is done by excavating *(trenching)* fresh exposures of a fault or finding natural exposures in canyon walls, and applying standard relative age dating techniques (please review the material of Exercise #7; the laws of superposition and crosscutting relationships will apply to **Problem #2** below). Once a relative age sequence is worked out, datable material is extracted from different rock layers and sent to the laboratory for geochronological analysis (note that finding datable material, usually charcoal or volcanic ash, is easier said than done!). Under ideal circumstances, these ages will closely constrain the time of fault movement. Commonly, one can only determine a broad time interval during which the fault could have been active. As might be expected, we worry most about faults that have moved during *Quaternary* time (the youngest period of the Cenozoic era; see **Figure 18.2**).

The Alquist Priolo Act (1978) is a California law setting guidelines for construction near faults that are deemed seismogenic. According to this law, developers may not position structures designed for human occupation within 50 feet of an active fault. An *"active fault"* is legally defined to be a fault that displaces *Holocene* deposits, i.e., Earth materials that are less than 11,700 years old. Along similar lines, a *"potentially active fault"* is one that displaces *Quaternary deposits* (sedimentary or volcanic materials dated at less than 2.59 million years before present). Further study in some areas may show that potentially active faults offset Holocene deposits; in these cases the fault must be reclassified as active. In event of this possibility, the law requires that detailed fault trenching studies be conducted within a zone 660 feet from a potentially active fault, to search for Holocene fault offsets. An *"inactive fault"* is defined as a fault that has not moved in the past 2.59 million years. In other words, an inactive fault will not displace Earth materials of Quaternary age. **Problem #2** will provide students practice in determining the degree of activity of various faults. **Problem #3** is a land zoning exercise that requires knowledge of the Alquist Priolo Act to map out specific areas of legally defined seismic hazard.

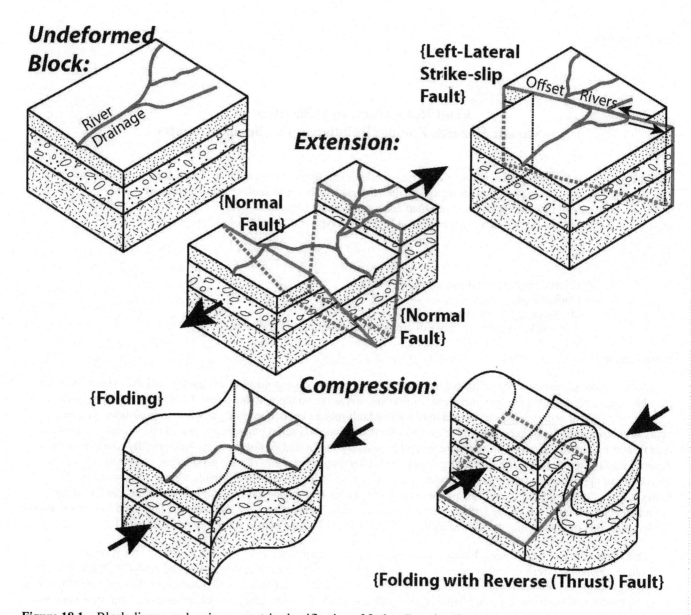

Undeformed Block:

River Drainage

Extension:

{Normal Fault}

{Normal Fault}

{Left-Lateral Strike-slip Fault}

Offset Rivers

Compression:

{Folding}

{Folding with Reverse (Thrust) Fault}

Figure 18.1. Block diagrams showing geometric classification of faults. Drawings by J. Nourse.

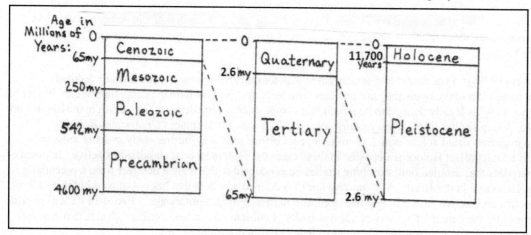

Age in Millions of Years:

	0		0		0
65my	Cenozoic		Quaternary	11,700 Years	Holocene
250my	Mesozoic	2.6my			
542my	Paleozoic		Tertiary		Pleistocene
4600my	Precambrian	65my		2.6 my	

Figure 18.2. A simplified version of the **Geologic Time Scale**.

The intensity of ground shaking during a given earthquake at a specific site will depend in general upon three parameters: (a) the Richter magnitude, (b) distance of the site from the earthquake source, and (c) the strength of foundation materials underlying the site. **Problems 5 and 6** explore aspects of parameter (c). One can expect more severe seismic shaking to take place in areas situated on the weakest foundations (**Figure 18.3**). The relative strengths of various Earth materials will depend predominantly on how the grains of the material are held together. Water content of porous materials may also be a factor, because water has no shear strength when subjected to seismic shaking. Your instructor will provide you with more details of the grain-to-grain textures of various materials, and how they relate to strength. For a given earthquake, one finds that consolidated bedrock shakes the least, whereas water saturated clays and silts shake the most violently. The seismic response of unconsolidated gravel and alluvium falls somewhere between that for bedrock and clay.

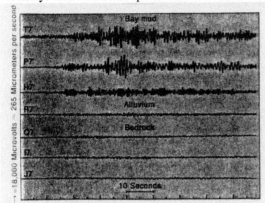

Recording of horizontal ground motion generated by an underground nuclear explosion in Nevada on sediment and bedrock in San Francisco. The effects are the same as those of an earthquake. The vertical scale shows the amount of ground motion; the horizontal scale shows elapsed time. Results are shown for three sites on recently deposited plastic mud, tens of meters thick, that contains more than 50% water; for one site underlain by clay, sand, and gravel hundreds of meters thick that contains less than 40% water; and for three sites underlain by semilithified and lithified sedimentary rocks of varied thicknesses.
Source: Borcherdt, 1975, p. 54, U.S. Geological Survey.

Figure 18.3. Effect of foundation type on the amplification of seismic surface waves and ground shaking.

Procedure / Problems

1. Fault-Controlled Valleys. In regions affected by active strike-slip faulting, rivers tend to erode along fault traces (this is because crushed rock associated with the faults tends to be more easily washed away than the surrounding rock). Certain streams in southern California provide striking examples of this process. Study the drainage map of the San Gabriel Mountains (**Figure 18.4**). Assuming that most of the rivers follow strike slip faults, can you deduce the main fault trends in this area? This may be accomplished by using different colors to outline drainages that flow in distinct directions.

A. Specify the main fault trends in terms of geographic directions, e.g., **SW, ESE, NNW**, etc.

B. Given what you know about strike-slip faults in southern California, state whether your faults are likely to record **right-lateral** or **left-lateral** displacements.

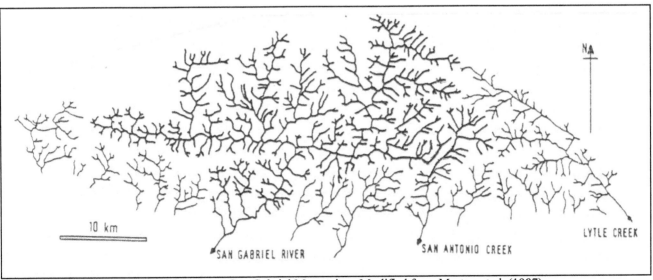

Figure 18.4. Drainage map of the eastern San Gabriel Mountains. Modified from Morton et al. (1987)

2. Age Constraints on Fault Activity. Study the illustrations of faults provided in **Figure 18.5**. Highlight each fault with a red pencil or pen. Then, applying the criteria introduced in **Exercise #6**, use the laws of superposition and crosscutting relationships to determine the <u>age constraints</u> for possible fault movement. For example, Fault X is younger than ??? years, and older than ??? years. Then, utilize the guidelines in the **Introduction section** of this chapter to classify each fault as one of the following: **legally active, potentially active, inactive,** or **indeterminate.*** *Hint:* The Bishop Tuff erupted ~760,000 years ago when Long Valley caldera formed. Mt. Mazama exploded in 5677 B. C. to create Crater Lake.

*For those faults marked indeterminate, please explain your reasoning. Certain faults may have insufficient age control to be classified precisely into one of the three categories.

Fault X

Fault Y

Fault A

Fault B

Fault 1

Fault 2

Fault 3

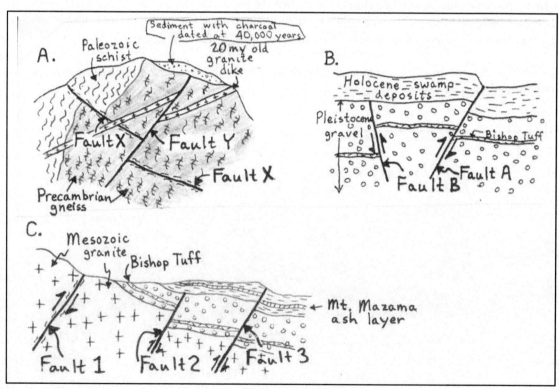

Figure 18.5. Cross sections illustrating hypothetical faults that record various movement histories. Drawings by J. Nourse.

3. Seismic Hazard Zonation. The property map below **(Figure 18.6)** shows surface ruptures associated with the 1992 Landers earthquake (M 7.4), located in California's Mojave Desert. Also mapped are previously recognized faults in the area (see legend). Following guidelines of the Alquist Priolo Act (**see Introduction section**), please zone the map area for purposes of residential development. Use a distinct color with a /xxxxxxxx\ pattern to indicate precisely those areas that \xxxxxxxx/

may **not** be built upon legally. Be sure to utilize the map scale.

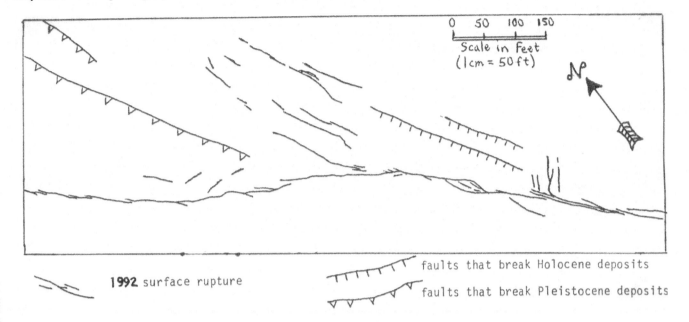

Figure 18.6. Map showing Quaternary fault scarps along one area of the Landers earthquake rupture zone.

4. Figure 18.7 (next page) is a map that zones land in the conterminous United States for potential seismic hazard. Notice that the risk zones have irregular or non-uniform shapes. Why is this? The basic answer is that earthquakes are not randomly distributed. This problem will explore the spatial relationships. You will also gain experience plotting locations on a map using a latitude-longitude grid.

    A. Refer to **Table 18.1** on **page 103**. Think about where these places are located. Then plot the **location** of each of the 20 earthquakes listed in the table onto the map of **Figure 18.7**. Use a **red star** to locate the epicenter of each earthquake. Also use a **black pen** to specify the **date** of each quake on the map, and its **magnitude**.

    B. Now inspect your completed earthquake epicenter map. It should be fairly obvious why the seismic risk is zoned as shown!

    C. Please list below some safety steps you should take (or research you might conduct) before purchasing a house and moving to one of the areas marked as Zone 3 (actually, you probably already live in one of these areas!!):

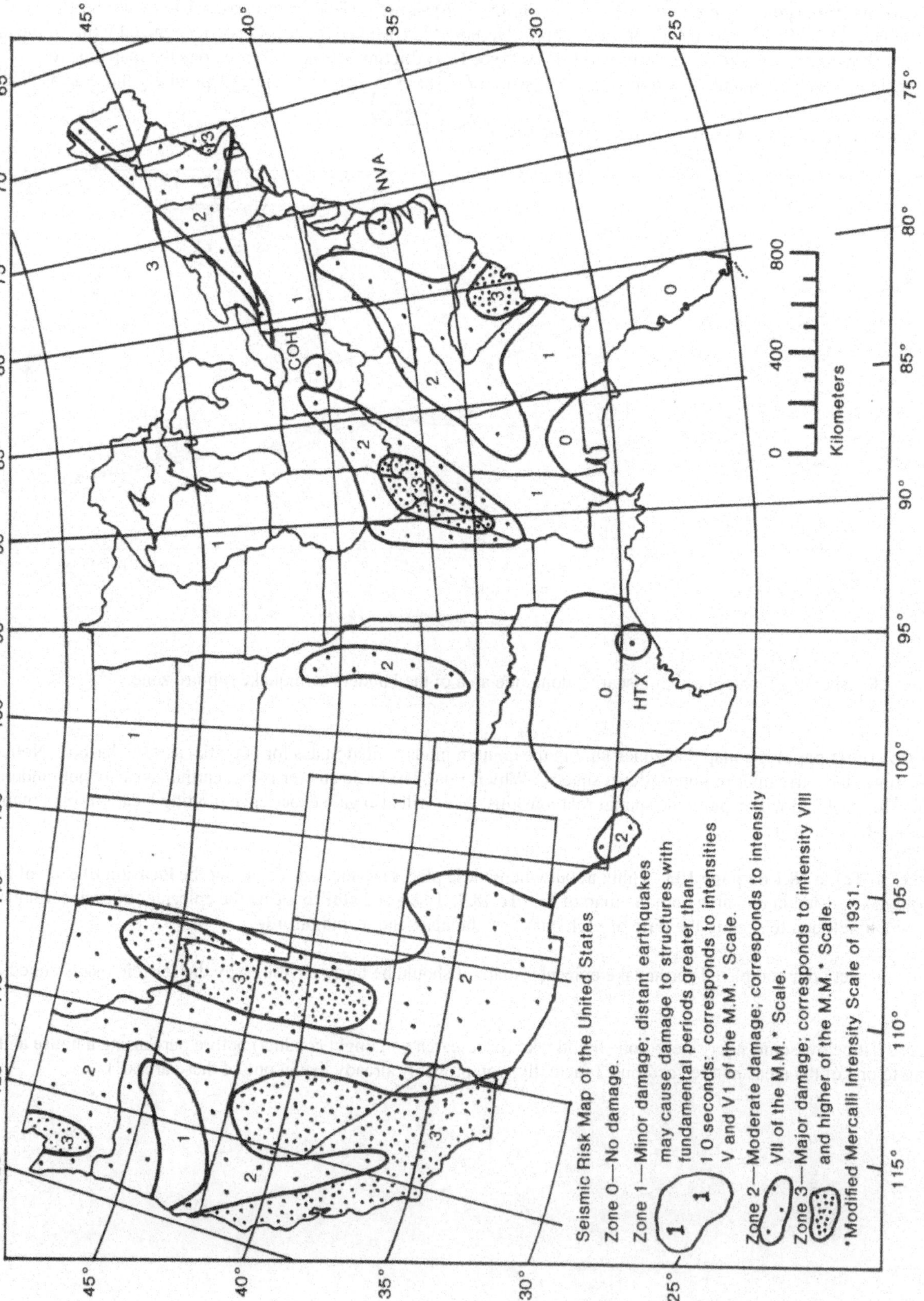

Seismic Risk Map of the United States
Zone 0—No damage.
Zone 1—Minor damage; distant earthquakes may cause damage to structures with fundamental periods greater than 1 0 seconds: corresponds to intensities V and V1 of the M.M.* Scale.
Zone 2—Moderate damage; corresponds to intensity VII of the M.M.* Scale.
Zone 3—Major damage; corresponds to intensity VIII and higher of the M.M.* Scale.
*Modified Mercalli Intensity Scale of 1931.
Kilometers
0
400
800
COH
NVA
HTX

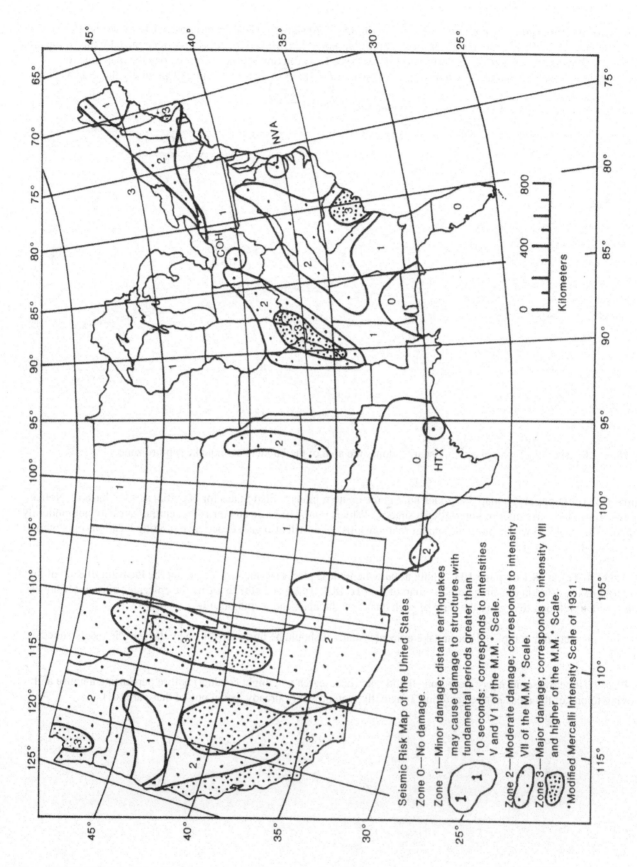

Figure 18.7. Seismic risk map of the lower forty eight United States. Numbers represent degree of risk: 0 for no risk at all to 3 for major damage expected from earthquakes. Modified from map by Ward's Natural Science Establishment.

Table 18.1   The Largest Earthquakes in the Contiguous United States

(compiled from wwwneic.cr.usgs.gov)

Rank	Location	Date	Magnitude	Latitude	Longitude
1	New Madrid, Arkansas	Dec. 16, 1811	8.1	35.6N	90.4W
2	New Madrid, Missouri	Feb. 7, 1812	~8.0	36.5N	89.6W
3	Fort Tejon, California	Jan. 9, 1857	7.9	35.7N	120.3W
4	New Madrid, Missouri	Jan. 23, 1812	7.8	36.3N	89.6W
5	Imperial Valley, California	Feb. 24, 1892	7.8	32.8N	115.6W
6	San Francisco, California	Apr. 18, 1906	7.8	37.7N	122.5W
7	Pleasant Valley, Nevada	Oct. 3, 1915	7.7	40.5N	117.5W
8	Owens Valley, California	Mar. 26, 1872	7.6	36.7N	118.1W
9	Trinidad, California	Nov. 8, 1980	7.4	41.1N	124.3W
10	Lake Chelan, Washington	Dec. 15, 1872	7.3	47.9N	120.3W
11	California-Oregon Coast	Nov. 23, 1873	7.3	42.0N	124.5W
12	Charleston, South Carolina	Sept. 1, 1886	7.3	32.9N	80.0W
13	West of Eureka, California	Jan. 31, 1922	7.3	41.0N	125.5W
14	Kern County, California	21-Jul-52	7.3	35.0N	119.0W
15	Hebgen Lake, Montana	Aug. 18, 1959	7.3	44.7N	111.2W
16	Landers, California	28-Jun-92	7.3	34.2N	116.4W
17	Borah Peak, Idaho	Oct. 28, 1983	7.3	44.0N	113.9W
18	Santa Barbara Channel	Dec. 21, 1812	7.1	34.2N	119.9W
19	Olympia, Washington	Apr. 13, 1949	7.1	47.0N	123.0W
20	Northridge, California	Jan. 17, 1994	6.8	34.2N	118.5W

5. Prediction of Ground Shaking Intensity for the next Earthquake

Robert Borcherdt (1975) conducted a detailed forensic study of the effects of ground shaking related to the 1906 (M= 7.8) San Francisco Earthquake. His US Geological Survey Professional Paper 941-A, entitled **"Studies for seismic zonation of the San Francisco Bay region"** is essential reading for anyone who lives in that region. Interestingly, Borcherdt's predictions were largely borne out during the 1989 (M= 7.1) Loma Prieta earthquake.

Below we will conduct a simple tracing activity to exemplify the relationship between natural foundation / ground strength (determined by surface geology) and the intensity of seismic shaking that might be expected in an earthquake-prone region

A. Place a piece of tracing paper or vellum over the seismic intensity map of **Figure 18.7** (next page). Then trace out and subdivide areas corresponding to different intensities of shaking during the 1906 San Francisco earthquake. Label your areas with 1's, 2's, or 3's, according to the legend on **Figure 18.7**.

B. Now place your tracing paper over the geologic map of **Figure 18.8**. Please describe the relationship between type of foundation material and intensity of ground shaking. Be sure to discuss the seismic response of each Earth material mapped. It may be helpful to refer to **Figure 18.3**. Where would you prefer to live if your residence for the next 20 years were to be San Francisco?

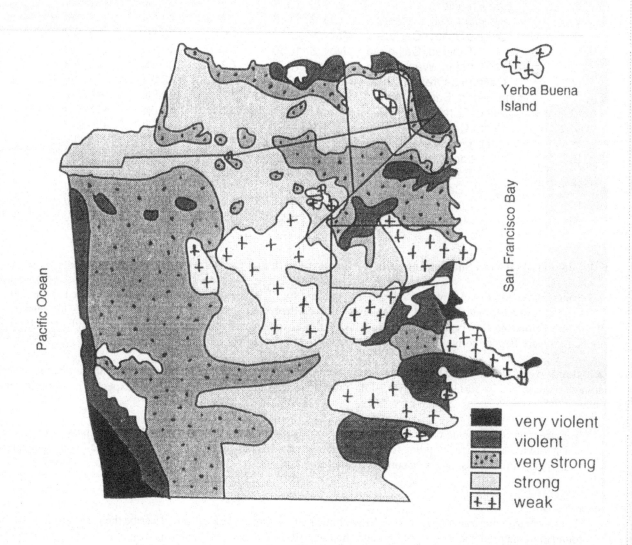

Figure 18.7. Map of San Francisco showing the relative intensity of ground shaking during the 1906 earthquake (modified from Borcherdt, 1975).

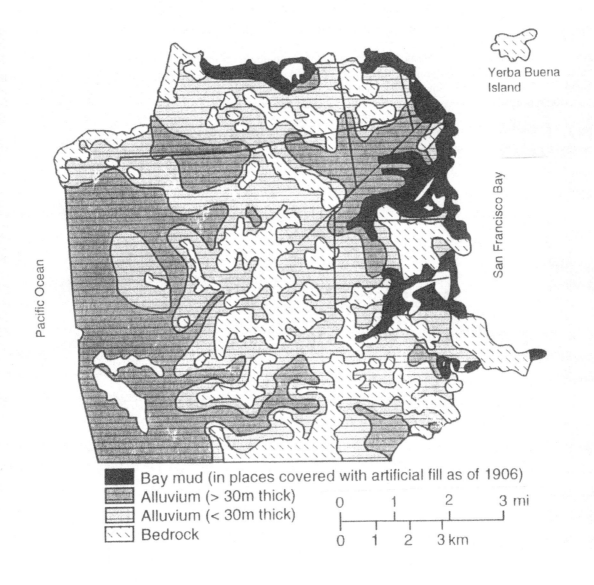

Figure 18.8. Generalized geologic map of San Francisco, showing various types of soil versus sedimentary or metamorphic bedrock (from Borcherdt, 1975).

6. Prediction of Seismic Shaking Effects for a Hypothetical Southern California Earthquake

Now let's apply what we have learned. As you may know, southern California residents have been forewarned about the potential for a large magnitude earthquake ('the Big One') on the San Andreas fault. According to some researchers, a magnitude 8+ earthquake might happen at any time. Below we will ponder the variation of ground shaking that might be expected during such an event, given the complex geology of the region.

Figure 18.9 below is a simplified geologic cross section showing the Los Angeles basin and adjacent San Gabriel Mountains. Note the San Andreas fault, where a hypothetical earthquake is generated.

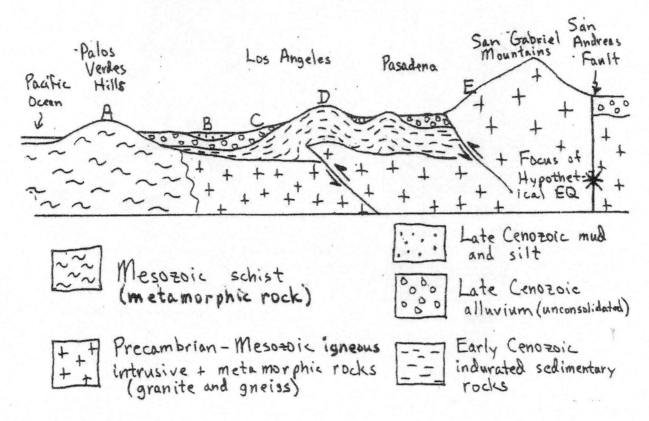

Figure 18.9. Generalized geologic cross section through the Los Angeles area. Note location of hypothetical earthquake. Drawing by J. Nourse.

Suppose a magnitude 8 earthquake were generated on the San Andreas Fault at the location indicated by the   *****
Considering distance from the earthquake and type of ground material, please rank the **intensity of ground shaking** one might expect at the five sites **A** through **E**. Use **1** for least intense shaking, and **5** for most intense shaking. Please also explain the reasoning for each of your choices:

Ground Site	Predicted Intensity of Ground Shaking (1 to 5)	Explanation (what is the logic behind your choices?)
A		
B		
C		
D		
E		

Exercise #19

Name: _____

Class Number: _____

Basic Seismology: Richter Magnitude and Earthquake Epicenter Location; Moment Magnitude and Ground Acceleration

Objective: Students will apply knowledge of seismic waves and seismographs to determine Richter magnitude and earthquake epicenter location; also moment magnitude and ground acceleration of hypothetical earhquakes.

Materials: compass (to draw circles)
ruler with a millimeter scale
calculator
pencil

Introduction

According to the plate-tectonic theory, the Earth's surface is divided into more than 20 moving plates. These plates can bump together, move apart, and grind past each other. The edges of these plates are outlined by volcanic activity and earthquakes (see also **Figure 1.1**).   When two rock masses move past each other, the sudden movement causes earthquakes. The place where this slippage actually occurs is the *focus* or *hypocenter* of the earthquake, and the location on the Earth's surface vertically above the focus is the *epicenter.* Your instructor will help you draw a simple picture to bring out this relationship. Earthquake epicenters are commonly plotted on maps like the one you created in **Exercise 18, Problem #4.**

Seismic Waves

When an earthquake happens, it creates several types of vibrations. These can be divided into body and surface waves; we will only be working with body waves today, but you should note that the surface waves usually cause most of the damage to structures from earthquake vibrations. The type of body wave that is transmitted fastest is a *compression wave;* the particles of rock vibrate in the same direction that the wave moves. This motion is similar to that of sound waves, with alternating expansion and compression. Like sound, these waves can be transmitted through solids, liquids, or gases. Compressional waves are also called push-pull, primary, or *P waves*; as you can see in **Figure 19.1**, these waves appear first on seismograms that record earthquake waves.

Another type of body wave is the *shear* (or transverse) *wave*, in which particles vibrate in right angles to the direction that the wave travels. The particles move in the same way a wave moves along a guitar string that is plucked. Because these are slower than P waves, they arrive at a recording station second. From this come the names secondary or *S waves*. After the S waves arrive, another type of seismic wave, the surface waves, arrive.

Because of the different velocities of the P and S waves, the S arrives at a seismograph station after the P wave. The *lag time* (difference in arrival times) can be used to calculate the distance to the epicenter. Lag time is usually measured in seconds, although minutes may be used for a distant earthquake.   From the difference in arrival times of the P wave and the S wave (the *lag time*), and average velocities for the P and S waves, *seismologists* can make the necessary calculations to determine the location of the epicenter and the time of the earthquake. **Problem #2** below provides illustrates the procedure.

The Richter Magnitude Scale

The magnitude of an earthquake can be measured both quantitatively by applying the *Richter Scale* and qualitatively by assessing the relative *intensity* of ground shaking. This exercise and **Exercise #18** explore both important aspects of earthquake magnitude. In **Problem #1** below you will have the opportunity to determine the Richter magnitude for several earthquakes. The Richter magnitude (M_L) was devised by Caltech seismologists Charles Richter and Beno Gutenburg in 1935 to measure the "size of an earthquake at its source." This is done quantitatively by measuring the maximum *amplitude* of the S wave on a seismograph situated 100 kilometers from the earthquake source. This amplitude is measured in millimeters (see **Figure 19.1**). An amplitude of 0.1 mm corresponds to a Richter magnitude of 2.0. One mm amplitude represents 3.0 on the Richter scale, 10 mm (of 1 cm) represents magnitude 4.0, 100 mm (or 10 cm) represents a magnitude 5.0, and so on. Unfortunately, most seismographs will not be located exactly 100 km from an earthquake. Thus, an adjustment must be made to account for different distances.

Figure 19.1 illustrates a graphical technique that takes variable distance into account. Remember that distance to an earthquake is proportional to the S-P lag time, which can be measured off any seismograph. You will use this drawing to determine Richter magnitudes from several different seismographs. Simply measure the S-P lag time (in seconds) off the seismograph, and plot a point on the left-hand scale on **Figure 19.1**. Then measure the maximum S-wave amplitude **in millimeters** and position a point on the right-hand scale. The straight line that connects these two points will intersect the scale for Richter magnitude. Read this value (to the closest one tenth) directly off the center scale.

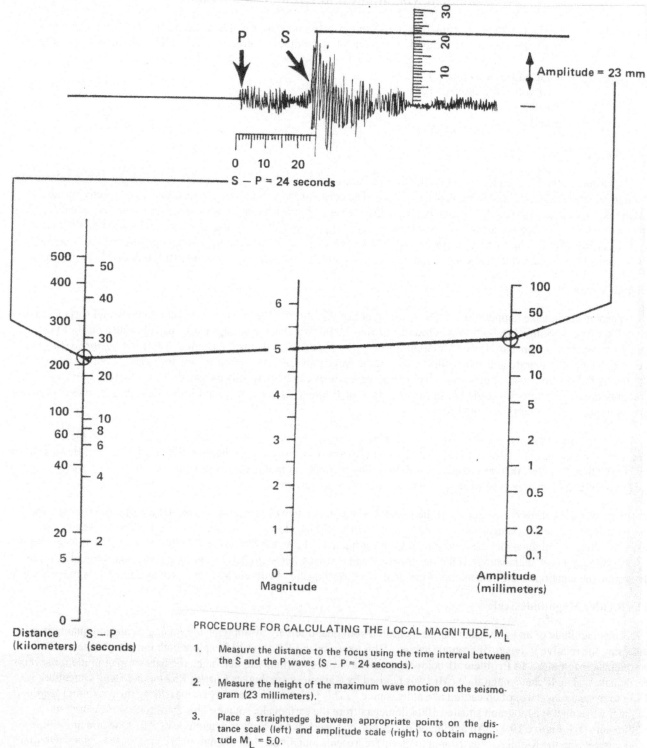

Figure 19.1. Graphical technique for determining the Richter magnitude of an earthquake (after Bolt, 1978).

Procedure / Problems

Problem 1. Study the four seismographs shown below in **Figure 19.2**. Seismograph #1, #2, and #4 record real earthquakes; Seismograph #3 is a fictitious earthquake:

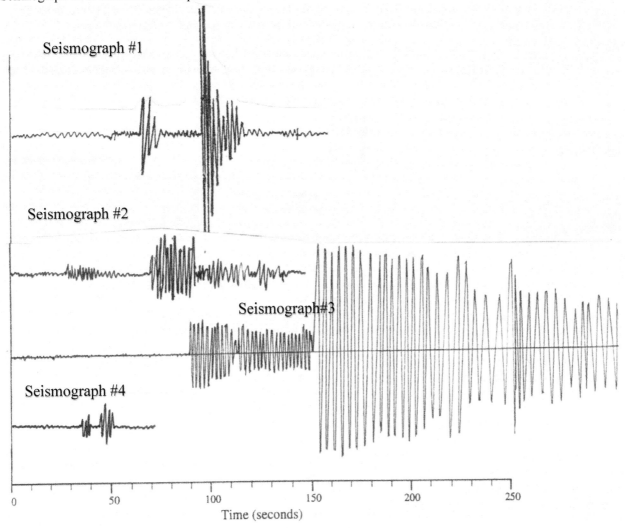

Figure 19.2. Seismographs to accompany Problem 1.

    A. Pick the first arrival times of the P and S waves,
    B. Determine the S-P lag time in seconds, and
    C. Measure the maximum amplitude of the S wave in millimeters. Then use **Figure 20.1** to determine the Richter magnitude of each earthquake. Fill out the chart below.

	S-P Lag Time (sec)	Distance (km)	Amplitude (mm)	Richter Magnitude
EQ #1				
EQ #2				
EQ #3				
EQ #4				

103

By performing the above procedures by hand, you have developed the basic understanding and skills needed to determine the _Richter magnitude_ of any earthquake. A different set of manual graphical procedures may be applied to locate the _epicenter_ of an earthquake. Imagine having to carry out these same procedures repetitively, day after day! Given the potential tedium of this job, as you might suspect, somebody has probably written a computer program to do all the calculations. In fact, many programs are available on the Internet for the public to utilize. With modern animation and color graphics, it can be very educational (and fun) to experiment with different earthquakes, and compare your results.

Problem 2. The web site below provides practical, user-friendly tutorials that will help you visualize the procedures involved with determining **Earthquake Epicenter Location** and **Richter Magnitude**. Simply follow the associated links and explore the various examples. One of the nice things about this site is the fact that real earthquake data is utilized. Thus, you will have an opportunity to learn about earthquakes in your own geographic region of interest.

http://www.sciencecourseware.org/VirtualEarthquake/

(These exercises were designed by **Dr Gary A. Novak** at Geology Professor at **California State University, Los Angeles**)

The Moment Magnitude Scale

It turns out that Richter magnitude (M_L) does not work well for measuring magnitude 7+ earthquakes. This is because nearby seismographs tend to get knocked off scale for large local events—an effect called "saturation". Also Richter's local magnitude scale is not very accurate at distances greater than 600km. So, the _Moment magnitude scale_ was developed in 1979 by Caltech seismologists Hiroo Kanamori and Thomas Hanks to allow measurement of large earthquakes and provide results consistent with the Richter scale for medium-sized earthquakes.

One advantage of Moment magnitude (M_w) is that it is based on physical parameters that measure energy release during fault rupture. The moment magnitude concept builds on theory proposed in 1972 by MIT geophysics professor Keiiti Aki. In simple terms, the energy released during an earthquake is proportional to: (1) the area of fault rupture, (2) the average displacement (slip) on the fault during the earthquake, and (3) the rigidity of the rock adjacent to the fault. In mathematical terms, seismic moment is define as follows:

$$M_o = (A)(D)(\mu), \qquad \text{where}$$

A = Area of fault rupture in $\mathbf{m^2}$, as measured by the distribution of the main earthquake and immediate aftershocks;
D = Displacement along the fault (in **m**) during its rupture, and
μ = Modulus of Rigidity or Shear Modulus of the local rock in **Pascals ($\mathbf{N/m^2}$)**. For igneous or metamorphic rocks at seismogenic depths, **μ** values typically range from 2 to 4 X 10^{10} Pascals

Note that the seismic moment calculation yields units of **Newton-meters** which is **force times distance**. Those of you who are engineers may recall that these are the same units as **energy**!

Now let's define the Moment Magnitude scale, $\mathbf{M_w}$, a dimensionless number. The "w" signifies that this represents the _mechanical work_ accomplished during faulting:

$$M_w = 2/3 \log_{10}(M_o) - 6.05$$

Because it is a measure of energy release, the moment magnitude scale allows one to compare the actual energy difference between two different earthquakes. The formula below accomplishes this:

$$\Delta E = 10^{1.5(m1-m2)}, \qquad \text{where}$$

ΔE = the difference in energy release,
m1 = moment magnitude of the larger earthquake, and
m2 = moment magnitude of the smaller earthquake

Applying this equation, it may be learned that a M = 7 earthquake is 31.6 times more energetic than a M = 6 earthquake. _Thus, an increase in one in Richter magnitude represents a 10x greater S-wave amplitude, but 32x greater energy release!_

104

Quantitative Prediction of Ground Acceleration

In areas where the near-surface geology is well-characterized, it is possible to predict the amount of *ground acceleration*, $a_{max/g}$, for an earthquake of a given *moment magnitude*. The calculated parameter $a_{max/g}$ is the fraction of Earth's gravitational acceleration, g, that the ground would move during an earthquake. A ground acceleration of 1 or greater theoretically would cause a building (or person) to be lifted off the ground surface! Procedures for predicting ground acceleration of the foundation, developed by Coduto et al., (2011), are very important for engineering design of buildings in earthquake-prone areas.

In the western United States where major contrasts exist between fault bounded bedrock mountain ranges and alluvial basins, the governing equation (developed by Boore et al., 2008) is:

$$\log (a_{max/g}) = -0.38 + 0.216(M_w-6) - 0.777\log R + 0.158G_B + 2.54G_C \text{ , where}$$

$a_{max/g}$	is the peak horizontal ground acceleration at the ground surface (as a fraction of g)
M_w	is the moment magnitude of the earthquake
$R = (d^2 + z^2)^{1/2}$	is the distance to the earthquake **hypocenter** (d = distance to epicenter; z = depth; both in meters)
G_B and G_C	are empirical coefficients (material constants) related to the type of ground, as defined below:

Site Class (Type of Surface Geology)	Seismic Shear Wave Velocity in Upper 30 m	G_B	G_C
Coherent igneous, sedimentary or metamorphic bedrock or	> 750 m/s (> 2500 ft/s)	0	0
Sand and gravel at low confining pressure	360 – 750 m/s (1200 – 2500 ft/s)	1	0
Water-saturated sand, silt, or mud at low confining pressure	180 – 360 m/s (600-1200 ft/s)	0	1

In the eastern United States, a different equation (developed by Toro et al. (1997) applies because the surface geology is dominated by coherent bedrock, with fewer deep alluvial basins:

$$\ln (a_{max/g})_{rock} = 2.20 + 0.81 (M_w-6) - 1.27\ln R + 0.11 \max \{\ln(R/100), 0\} - 0.0021R$$

To summarize, it should be possible to predict the ground acceleration at a given site if a reasonable estimate of maximum credible earthquake magnitude (Mw) is known, and the details of the site geology are well-constrained. Let's try out the procedure by analyzing a hypothetical earthquake in southern Mexico.

Problem 3. Figure 19.3 (next page) shows a hypothetical earthquake generated by the subduction zone of the coast of southern Mexico. The epicenter of the main shock is indicated with a large star ☆ . Aftershocks that delimit the extent of fault rupture are shown with small x's. A cross section through the area is illustrated in **Figure 19.4**. In this example, we will evaluate the ground response at four sites labeled **A, B, C,** and **D. Site A** is a granite bedrock exposure in Acapulco. **Site B** is sandy-gravel alluvium in a small coastal valley. **Site C** is a lake-bed deposit in the old part of Mexico City, composed of water-saturated silt and mud. **Site D** is an andesite bedrock exposure near Mexico City.

Now we will determine the moment magnitude of this earthquake and predict the site responses at the four sites **A, B, C,** and **D.** Please apply the appropriate equations from **pages 110-111,** following the hints given below.

A. Calculate the Seismic Moment, M_o, given that average displacement (slip) on the fault is **10 m**, and the modulus of rigidity is **3 x 10^10 N/m²**. *Hint:* Determine the area of fault rupture by measuring the length of the aftershocks on **Figure 19.3,** and the width of the aftershock zone on **Figure 19.4.** Express your answer in Newton-meters. Show all work below:

B. Calculate the Moment Magnitude, M_w, for this earthquake. Show your work below:

C. Use the equation applicable to the western United states (**see page 111**), to calculate the ground acceleration at each of the four sites shown on the map. You will need to use an appropriate empirical coefficient (G_B or G_C) for the ground type or surface geology. Show all relevant work below:

    Site A:

    Site B:

    Site C:

    Site D:

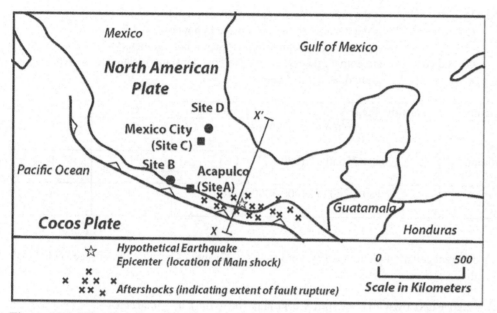

Figure 19.3. Map of southern Mexico showing location of the main shock and aftershocks associated with a hypothetical earthquake. The aftershock distribution indicates the length of the fault plane that ruptured. Note the location of four sites, A, B, C, and D with geologic character described in the problem statement above. The line **X-X'** marks the cross section of **Figure 19.4.** Drawing by J. Nourse.

X   X'

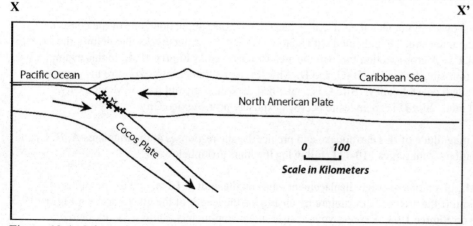

Figure 19.4. Schematic cross section through the ruptured subduction zone fault in southern Mexico. Drawing by J. Nourse.

Exercise #20

Name: _____

Class Number: _____

Neotectonics: Fault Slip Rates and Earthquake Recurrence Intervals

Objectives: To develop an understanding of how geologists calculate slip rates and earthquake recurrence intervals for active faults.

Materials: pencil
ruler
calculator

Introduction

The San Andreas Fault marks the primary boundary between the North American and Pacific tectonic plates (**Figure 20.1**). This major active fault extends for about 1100 kilometers (700 miles) from northern to southern California, forming a **transform plate boundary** along which the two sides move laterally (sideways) past each other. If you were standing on the North American Plate looking out towards the Pacific Ocean, the Pacific Plate would be moving to your right. The San Andreas fault, therefore, is a **right-lateral strike-slip fault**, as indicated by the large arrows in **Figure 20.1**. This sense of movement is opposite that shown by the left-lateral fault of **Figure 18.1**.

The Carrizo Plain in central California (**Figure 20.1**) is a famous location for observing the geomorphology of the San Andreas Fault (e.g., Sieh and Jahns, 1984; Sieh and Wallace, 1987). Here, the fault cuts a prominent linear trace through an area of arid hills and meadows, offsetting stream channels and creating landforms such as fault scarps, sag ponds, and shutter ridges (**Figure 20.2**). At Wallace Creek on the Carrizo Plain, right-lateral fault movement over the past several thousand years has offset a series of stream channels that cross the fault trace. In this exercise, you will use a LiDAR map of Wallace Creek (**Figure 20.3**) to measure fault offsets and calculate the average slip rate for the San Andreas Fault. Using the slip rate and earthquake slip data for the 1857 M8.0 Fort Tejon earthquake, you will calculate the recurrence interval (repeat time) for major earthquakes along this part of the San Andreas Fault.

Figure 20.1. Map of the San Andreas Fault plate boundary in California showing the location of Wallace Creek (from SCEC, 2008). The 1906 and 1857 earthquake rupture zones are shown as solid lines along the fault. Dashed lines indicate creeping fault segments or segments without historic ruptures. Large arrows show relative movements of the North American and Pacific tectonic plates on either side of the fault.

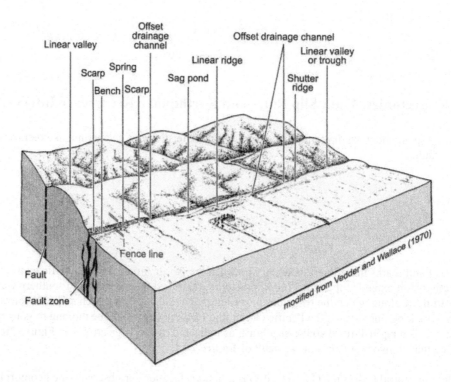

Figure 20.2. Block diagram of typical landforms developed by lateral fault movement along a strike-slip fault. Modified from Vedder and Wallace, 1970.

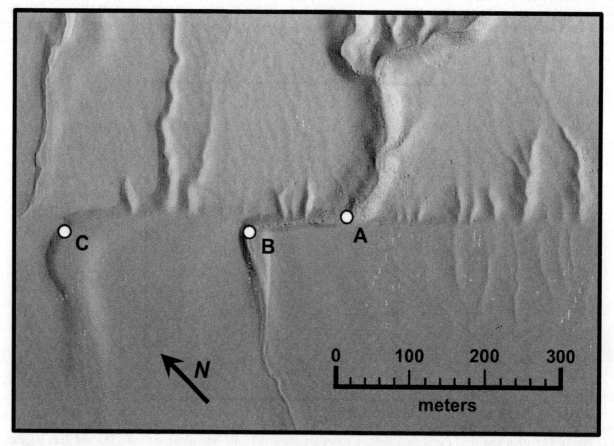

Figure 20.3. Airborne LiDAR Digital Elevation Model (DEM) of the San Andreas Fault at Wallace Creek, Carrizo Plain, California. Modified from Open Topography, 2012.

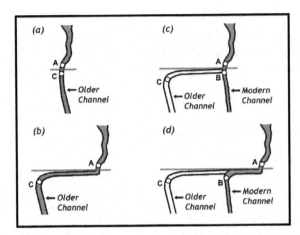

Figure 20.4. Four time steps in the long-term evolution of Wallace Creek at the San Andreas Fault: **a)** when the older channel was first formed, flowing straight across the fault; **b)** after the older channel has been offset along the fault, but before the newer modern channel was formed; **c)** just after the older channel was abandoned and the modern channel was formed, flowing straight across the fault; **d)** in its present form, after the modern channel has been offset along the fault. The straight horizontal line marks the fault trace. The shaded stream channel marks the active channel during each time step. Points A, B, and C correspond to the same points shown on the LiDAR map in **Figure 20.3**. Modified from SCEC, 2008.

Procedure / Problems

Problem 1. Measuring Fault Offsets and Calculating Slip Rates

Using the LiDAR map of Wallace Creek (**Figure 20.3**), answer the following questions:

A. Use a ruler and the scale on the map to measure the linear distance between the lower end of the modern upstream channel of Wallace Creek (Point A) and the upper end of the modern offset channel (Point B). Give your answers in meters (m) and in feet (ft). Show all calculations at the right. Report results to nearest "tenth" (one place after decimal).

      Distance from A to B = _____ m
      Distance from A to B = _____ ft

B. Use a ruler and the scale on the map to measure the linear distance between the lower end of the modern upstream channel of Wallace Creek (Point A) and the upper end of the older beheaded channel (Point C). Give your answers in meters (m) and in feet (ft). Show all calculations at the right. Report results to nearest "tenth" (one place after decimal).

      Distance from A to C = _____ m
      Distance from A to C = _____ ft

C. Radiocarbon dating of charcoal samples collected at Point C show that the most recent (youngest) stream deposits in the older beheaded channel are approximately 3800 years old. This indicates that the beheaded channel was abandoned at this time and that stream flow jumped to the new modern channel (Point B) straight across the fault (**Figure 20.4**, Time Step "c").

Since that time (3800 years ago), fault motion has separated Point A and Point B by the amount that you measured in **Question #1A** (**Figure 20.4**, Time Step "d"). Your measured distance represents the amount of "fault offset" or "slip" that occurred during the time interval between 3800 years ago and today (when the LiDAR map was created).

Based on your measurement of fault offset between Point A and Point B, calculate the average "slip rate" along the San Andreas Fault at this location. Give your results in meters per thousand years (m/ky), millimeters per year (mm/yr), feet per thousand years (ft/ky), and inches per year (in/yr). Show all calculations at the right. Report results to nearest "tenth" (one place after decimal).

      Slip Rate = _____ m/ky
      Slip Rate = _____ mm/yr
      Slip Rate = _____ ft/ky
      Slip Rate = _____ in/yr

D. Using the slip rate that you calculated in **Question #1C** and the offset between Point A and Point C that you measured in **Question #1B**, calculate the age of formation of the older beheaded channel at Point C . In other words, determine how much time has passed since Point C was lined up with Point A on opposite sides of the fault (**Figure 20.4**, Time Step "a").

Formation Age of Older Beheaded Channel at Point C = _____ years

Problem 2. Estimating Earthquake Recurrence Intervals

Along some sections of active faults, slip occurs continuously and gradually in a process referred to as "fault creep". However, along most segments of active faults, the two sides of the fault remain stuck together due to friction between the rocks on either side. In this case, "strain" builds up over time as the two sides of the fault try to move, but cannot. However, once the built-up strain overcomes the frictional resistance, the fault will move in a sudden burst, releasing stored seismic energy in an earthquake. This type of fault behavior is referred to as "stick-slip". The total slip that occurs during the earthquake represents the amount of accumulated strain since the last earthquake (as long as it released all of the built-up strain). Therefore, we can use the average long-term slip rate for a fault to make inferences about how often large earthquakes occur and when the next one is likely to occur.

On January 9, 1857, California was rocked by a major earthquake centered on the central segment of the San Andreas Fault (**Figure 20.1**). Known as the Fort Tejon Earthquake, this event ruptured the ground surface along 360 kilometers (225 miles) of fault line, and produced strong shaking that was felt as far away as Sacramento and Tijuana (**Figure 20.5**). While the ground shaking caused substantial damage to some adobe brick buildings, only two people lost their lives in what was then a sparsely populated region. With an estimated Richter Magnitude of M8.0, this event represents the legendary "Big One", the largest probable earthquake that the San Andreas Fault can produce. When the next "Big One" strikes in this now heavily populated region, there will be significant potential for death, injury, and destruction. It is important, therefore, for geologists to answer two important questions: 1) How often do major earthquakes like the Big One occur?, and 2) When will the next Big One strike in this area?

Figure 20.5. Isoseismal map of the M8.0 1857 Fort Tejon Earthquake (modified from Stover and Coffman, 1993) showing Mercalli Intensity values for ground shaking.

In the Carrizo Plain, the 1857 earthquake produced stream channel offsets of about 9 meters (30 feet). Fault trenching studies indicate that two earlier pre-historic earthquakes moved the fault by a similar amount. If we make the general assumption that the 1857 earthquake is typical of major earthquakes along this fault segment, we can calculate the average recurrence interval (repeat time) for these events and estimate the timing of the next event. This calculation requires knowing the average long-term slip rate for the fault and the average slip that takes place during each earthquake. In reality, we now understand that large earthquakes do not necessarily repeat at regular intervals. However, calculating average recurrence intervals for "characteristic earthquakes" is a useful starting point for evaluating earthquake hazards along any active fault.

Using the fault slip data that you calculated in **Problem 1**, answer the following questions:

A. Knowing that the 1857 earthquake produced 9 meters (30 feet) of fault slip at Wallace Creek, use the average slip rate you determined in **Question #1C** to calculate the average recurrence interval (repeat time) for an 1857-type earthquake. Show all calculations. Report results to nearest "tenth" (one place after decimal).

Average Recurrence Interval for the 1857 Earthquake = _____ years

B. How many years have passed since the 1857 earthquake? Based on your average recurrence interval, when should we expect the next M8.0 earthquake on the central San Andreas Fault? (Note: Due to the many uncertainties involved, this calculation is a "forecast" and not a "prediction".)

Time since the 1857 Earthquake = _____ years

Estimated year for the next M8.0 earthquake = _____

C. Based on the time since the 1857 earthquake, calculate the amount of strain that has accumulated on the fault since the earthquake (measured as a slip distance). This amount represents "missing slip" that eventually will be released in a future earthquake. Report your answers in meters and feet. Show all calculations. Report results to nearest "tenth" (one place after decimal).

Strain accumulated since the 1857 Earthquake = _____ meters
Strain accumulated since the 1857 Earthquake = _____ feet

D. Considering that a single M8.0 earthquake releases a total of 9 meters (30 feet) of strain, use your results from **Question #2C** to calculate the percentage of strain that has accumulated since 1857 toward producing the next earthquake?

Percentage of full earthquake slip accumulate since 1857 = _____ %

F. Based on 9 meters (30 feet) of earthquake slip, calculate how many 1857-type earthquakes it took to create the offset you measured earlier between Point A and Point B? Do the same for Point A and Point C. Show all calculations. (Note: Your answer may not be a perfect integer, but is likely to include a decimal fraction). Report results to nearest "tenth" (one place after decimal).

Number of earthquakes represented by offset between A and B = _____
Number of earthquakes represented by offset between A and C = _____

G. Why do you think that your answers to **Question #2F** are not perfect integers (an exact number of earthquakes), but rather include decimal fractions?

H. Considering that the fault has not moved since 1857, this means that your original slip rate calculations did not include the "missing slip" between 1857 and today. Do you think that this would have a significant impact on your slip rate calculation? Why or why not? What are some of the other uncertainties involved in your original slip rate calculation?

111

Exercise #21

Name: _____

Class Number: _____

Neotectonics: Marine Terrace Uplift

Objectives: To develop an understanding of how geologists use marine terraces to calculate tectonic uplift rates along coastlines.

Materials: colored pencils: black, yellow, orange, red, green, blue
ruler
calculator

Introduction

Marine terraces are a common landform along active tectonic coastlines around the world (**Figure 21.1**). These flat, stair-step surfaces represent ancient shorelines that have been lifted above the ocean by tectonic movements along fault lines. Through the study of marine terraces, geologists are able to calculate coastal uplift rates, measure spatial and temporal variations in tectonic deformation, and estimate recurrence intervals for earthquakes on coastal faults that generate vertical movement (uplift).

Figure 21.1. Uplifted late Pleistocene marine terraces at Wilder Ranch near Santa Cruz, California. Photograph by Jeff Marshall.

Wilder Ranch near Santa Cruz, California (**Figure 21.2**) is a famous location for observing the geomorphology of uplifted marine terraces (e.g., Bradley and Griggs, 1976; Hanks et al., 1984; Anderson, 1990; Weber, 1990; Anderson and Menking, 1994; Perg et al., 2001). Here, a flight of five prominent marine terraces step upward out of the ocean like a giant staircase. (**Figure 21.1**). These terraces are located just west of the San Andreas Fault and are thought to be the result of tectonic uplift generated by repeated earthquakes within a compressional bend along the fault. In this exercise, you will use a topographic map of the Wilder Ranch coastline to map the extent of a flight of five marine terraces. You will then create a topographic profile across the terraces to determine their elevations. Based on the terrace elevations, you will estimate their ages and calculate an average uplift rate by correlating your terraces with a global sea-level curve. Using the uplift rate and coastal uplift data from the 1989 M6.9 Loma Prieta earthquake, you will calculate the recurrence interval (repeat time) for similar earthquakes along this part of the San Andreas Fault.

113

Figure 21.2. Map of the San Andreas Fault plate boundary in California showing the location of the Wilder Ranch marine terraces near Santa Cruz. The 1906 and 1857 earthquake rupture zones are shown as solid lines along the fault. Dashed lines indicate creeping fault segments or segments without historic ruptures. Large arrows show relative movements of the North American and Pacific tectonic plates on either side of the fault. Map modified from Southern California Earthquake Center.

Marine terraces form by wave erosion of the rocks exposed along a coastline. This process creates a distinctive geomorphology with a variety of important features shown in **Figure 21.3**. As waves cut further into the land, they bevel off a relatively flat surface referred to as a **wavecut platform** (or intertidal platform). At the back of the wavecut platform, is a **seacliff**, which retreats over time as wave erosion cuts further into the landscape. The angle of intersection between the flat wavecut platform and the steep seacliff is referred to as the **shoreline angle** (or inner edge). In three dimensions, the shoreline angle forms a line that is parallel with sea level at the time of terrace formation. The shoreline angle, therefore, is a very important feature that is directly linked to past sea level.

As gradual tectonic uplift raises a terrace above sea level, the old wavecut platform and seacliff are abandoned. The platform will become covered with sediments, referred to as **marine terrace deposits**, consisting of beach sediments, stream deposits, and eroded colluvium from the adjacent seacliff **(Figure 21.3)**. The upper surface of the terrace deposits is referred to as the **terrace tread**, and the old seacliff at the back edge of the terrace is referred to as the **terrace riser**. The old, original shoreline angle (inner edge) lies buried beneath the terrace cover deposits at the intersection of the old wavecut platform and old seacliff.

The long-term interaction between rising and falling sea level, plus gradual tectonic uplift **(Figure 21.4)**, results in the formation of multiple marine terraces that rise out of the ocean in a stair-stepping sequence (or terrace flight). Over the past two million years of Earth's history **(Quaternary Period)**, global climate has fluctuated rapidly between cold periods **(stadials)** and warm periods **(interstadials)**. During the cold periods ("ice ages"), massive ice sheets formed on the continents, removing large volumes of water from the ocean and causing global sea level to drop. During the intervening warm periods, the ice melted and water returned to the ocean, leading to sea level rise. Using a variety of lines of evidence, oceanographers and geologists have constructed global **(eustatic)** sea level curves that show sea level fluctuations through time **(Figure 21.4)**.

The high and low stands of global sea level are numbered in increasing order, starting with our modern **(Holocene)** sea level high stand, which is referred to as OIS 1. The abbreviation **OIS** stands for **Oxygen Isotope Stage** which is a system, based on deep marine sediment chemistry, for distinguishing between the cold stadials and warm interstadials. With OIS 1 marking our modern warm high stand, OIS 2 refers to the Last Glacial Maximum low stand that occurred 18,000 years ago. Following this trend, odd OIS numbers refer to sea level high stands and even OIS numbers refer to sea level low stands **(Figure 21.4)**.

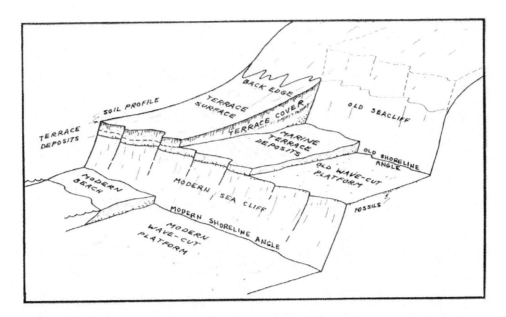

Figure 21.3. Block diagram with cut-away views showing key features of uplifted marine terraces. (From Weber and Allwardt, 2001).

Because marine terraces are formed by wave erosion cutting into the land, they are directly linked to rising sea level. This means that the old shoreline angle, at the base of an abandoned seacliff, marks the furthest inland advance of sea level during an interstadial high stand. Therefore, we can correlate every marine terrace in a flight of terraces directly to a unique sea level high stand. This is known as **sea level curve correlation** and provides a means of calculating the **terrace uplift rate**. Using a sea level curve correlation diagram **(Figure 21.4)**, geologists match each terrace to the sea level high stand during which the terrace was cut. A set of parallel lines are drawn that link the modern elevation of each terrace to the top of a high stand on the sea level curve. The slope of these lines (rise/run) is equal to the uplift rate (elevation/time). The lines are parallel because the uplift rate is the same for all terraces at this site (and is assumed to be constant over time). If the line for one or more terraces in a sequence does not intersect a high stand, then the correlation scenario (and uplift rate) being tested is incorrect.

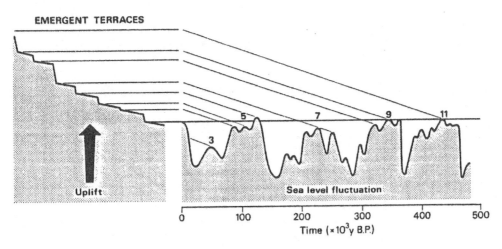

Figure 21.4. Example sea level curve correlation diagram for uplifted marine terraces (from Lajoie, 1986). A flight of uplifted terraces is shown at left. Horizontal lines mark the shoreline angle elevations above modern sea level (thick horizontal line). At right, the global sea level curve for the last 500,000 years is shown as a jagged line. Sea level high-stands (interstadials) are marked by odd numbers (OIS 1-11). The intervening low-stands (stadials) correspond with major glacial advances (ice ages). The sloping straight lines connect marine terrace inner edge elevations to the unique sea level high stand during which they formed. The slope of these lines (rise/run) is equal to the tectonic uplift rate (elevation/time) for this location.

115

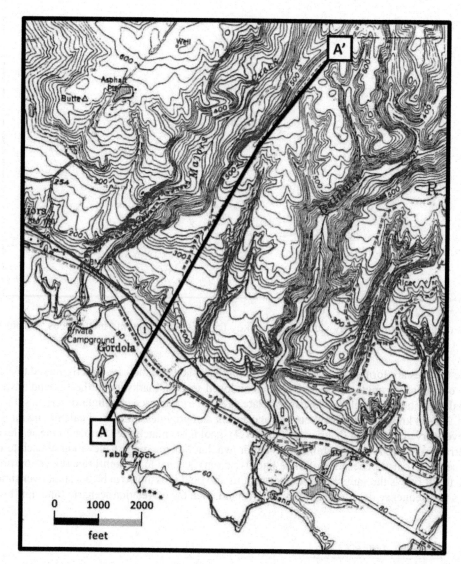

Figure 21.5. Topographic map of marine terraces at Wilder Ranch, near Santa Cruz, California. Line A-A' shows the topographic profile location. Elevation contour interval = 20 feet. Length scale at lower left.

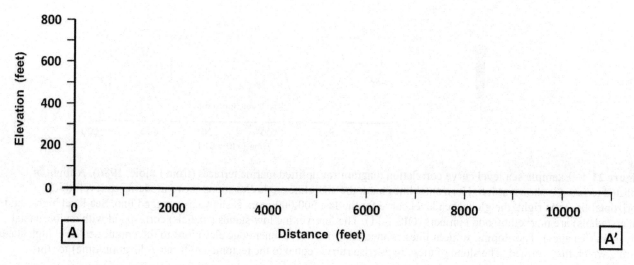

Figure 21.6. Graph for plotting topographic profile across the five marine terraces at Wilder Ranch. See topographic map in **Figure 21.5** for location of profile line.

116

Procedure / Problems

Part A. Mapping and Profiling Marine Terraces

Using the topographic map of marine terraces at Wilder Ranch (**Figure 21.5**), complete the following tasks:

1. On **Figure 21.5**, create a geologic map of the five marine terraces located between Majors Creek and Baldwin Creek. Label each of the terrace treads in the sequence as T1 (lowest) through T5 (highest). Use colored pencils to lightly shade the terrace treads (areas of relatively flat topography, where contour lines are far apart). Use the following colors: T1 (yellow), T2 (orange), T3 (red), T4 (green), and T5 (blue). Do not color the areas of steep topography (where contour lines are close together) that correspond with terrace risers (seacliffs), stream gullies, and canyons. (Hint: The T1 tread lies mostly between the 60 ft and 120 ft contour lines.)

2. Using the graph in **Figure 21.6**, create a topographic profile of the five marine terraces along transect line A-A' (shown on **Figure 21.5**). Use a ruler, and the distance scale on the map, to measure linear distances along the transect line. Mark a point on the map at each location where the transect line crosses an elevation contour line. Put a corresponding dot on the graph that matches the distance (x-axis) and elevation (y-axis) of each of these points. After graphing all of the points, draw a line that connects the dots. The resulting figure is a topographic profile showing the five marine terraces, including relatively flat treads and steep risers.

3. Use your topographic profile (**Figure 21.6**) to determine the shoreline angle elevations for the five marine terraces. Recall that the shoreline angle lies at the intersection of the paleo-wavecut platform and paleo-seacliff. To find the shoreline angle, use a ruler to draw a straight line along the terrace tread that extends beneath the terrace riser (cliff). Draw another straight line down the riser until it intersects the tread line you just drew. Record the elevation of the point of intersection. Because the tread lies atop the terrace cover deposits, you must subtract the thickness of the terrace deposits from the tread elevation to determine the true shoreline angle elevation. For the Wilder Ranch terraces, we will estimate the thickness of all terrace cover deposits at 10 feet.

Shoreline Angle Elevation for T1 = _____ ft

Shoreline Angle Elevation for T2 = _____ ft

Shoreline Angle Elevation for T3 = _____ ft

Shoreline Angle Elevation for T4 = _____ ft

Shoreline Angle Elevation for T5 = _____ ft

Part B. Sea Level Curve Correlations and Calculating Uplift Rates

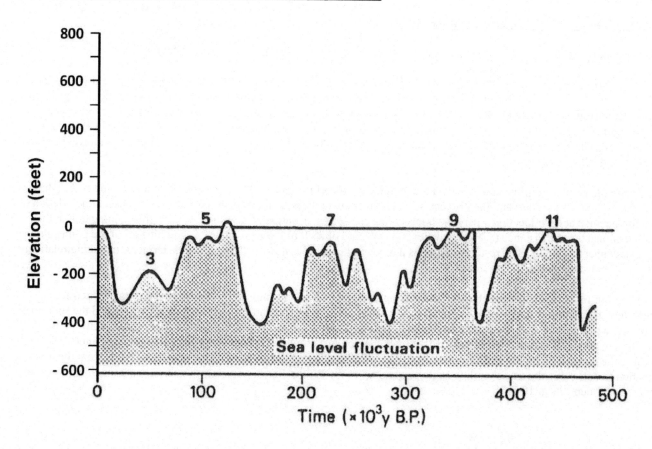

Figure 21.7. Sea level curve correlation diagram. Use this diagram to plot terrace shorline angle elevations (on y-axis) and link them with straight parallel lines to their corresponding sea level high stand. The slope of these lines (rise/run) equals the uplift rate (elevation/time). See **Figure 21.4** for example.

Using your topographic profile (**Figure 21.6**) and shoreline angle elevations, complete the following tasks:

1. Use the graph in **Figure 21.7** to correlate marine terrace elevations with high stands of the global sea level curve. On the y-axis (elevation), mark a dot corresponding to the shoreline angle elevations of each of the five terraces. Label the dots T1-T5.

2. Cosmogenic radionuclide dating of terrace deposits (Perg et al., 2001) indicates that T1 was formed during the OIS 3 sea level high stand. Use a ruler to draw a line connecting the elevation point for T1 to the top of the OIS 3 high stand on the sea level curve. Now draw a set of parallel lines from the elevation points of T2-T5 down to the sea level curve. If the correlation works correctly, all of these parallel lines should intersect a sea level high stand (or come close to one).

3. Did the correlation work? There are a number of variables at play, including your elevation measurements, thickness of terrace deposits, the accuracy of the sea level curve, and the accuracy of the cosmogenic age data. We also made an assumption that the uplift rate has remained constant through time. If your correlation did not work out, discuss here why you think it did not.

118

4. Based on the slope of the correlation lines in your plot (**Figure 21.7**), calculate the average uplift rate for the Wilder Ranch marine terraces. Recall that the slope of the lines (rise/run) is equal to the uplift rate (elevation/time). Note: If your correlations didn't work, use the line for T1 to calculate the uplift rate. Report your answer in feet per thousand years (ft/ky), inches per year (in/yr), meters per thousand years (m/ky), and millimeters per year (mm/yr). Please each calculation or unit conversion in the space above the corresponding question:

Average uplift rate for Wilder Ranch marine terraces = _____ ft/ky

Average uplift rate for Wilder Ranch marine terraces = _____ in/yr

Average uplift rate for Wilder Ranch marine terraces = _____ m/ky

Average uplift rate for Wilder Ranch marine terraces = _____ mm/yr

Part C. Estimating Earthquake Recurrence Intervals

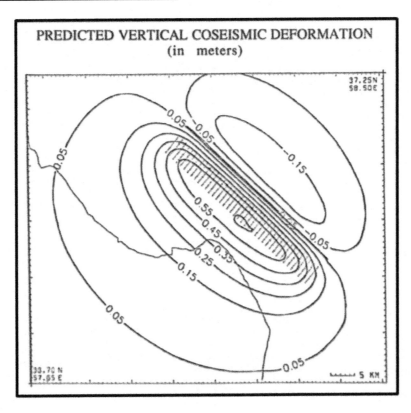

Figure 21.8. Fault dislocation model for 1989 Loma Prieta Earthquake showing contours of estimated coseismic uplift and subsidence (Modified from Anderson, 1990). Uplift of about 0.1 meters (0.3 feet) is estimated for the coastal area near the Wilder Ranch marine terraces.

On October 17, 1989, northern California was shaken by a Magnitude 6.9 earthquake centered southeast of Santa Cruz on the San Andreas Fault. Referred to as the Loma Prieta Earthquake, this event ruptured a locked segment of the fault deep beneath the Santa Cruz Mountains. Slip on the fault was oblique, with 5.6 feet (1.7 m) of horizontal motion, and 4.0 feet (1.2 m) of vertical motion, moving the Pacific Plate northwest and up relative to the North American Plate. Mathematical models (Anderson, 1990) show that the Loma Prieta Earthquake produced about 0.3 feet (0.1 meters) of uplift along the coastline at Wilder Ranch (**Figure 21.8**).

Using the average uplift rate that you calculated in Part B, complete the following tasks:

1.   Knowing that the 1989 earthquake produced 0.3 feet of uplift at Wilder Ranch (**Figure 21.8**), use the average uplift rate you determined in Part B to calculate the average recurrence interval (repeat time) for a Loma Prieta type earthquake. Show all calculations in the space below. Report results to nearest "tenth" (one place after decimal):

Average Recurrence Interval for the Loma Prieta Earthquake = _____ years

2.   How many years have passed since the 1989 earthquake? Based on your average recurrence interval, when should we expect the next Loma Prieta type earthquake in this area? (Note: Due to the many uncertainties involved, this calculation is a "forecast" and not a "prediction".)

Time since the 1989 Earthquake = _____ years

Estimated year for the next Loma Prieta earthquake = _____

3.   Based on 0.3 feet of earthquake uplift, calculate how many 1989-type earthquakes it took to uplift terrace T1 to its present elevation. Important: Note that T1 formed at the OIS 3 high stand which is below modern sea level. To get the total uplift of T1, add the elevation of T1 above modern sea level to the depth of OIS 3 below modern sea level. Repeat this same calculation for terraces T3 and T5. Show all calculations in the space above the corresponding question. Report results to nearest "tenth" (one place after decimal):

Number of earthquakes represented by uplifted terrace T1 = _____

Number of earthquakes represented by uplifted terrace T3 = _____

Number of earthquakes represented by uplifted terrace T5 = _____

4.   Do you think that it is valid to estimate earthquake repeat times using an uplift rate that is averaged over such a long period of time (10s to 100s of thousands of years)? Why or why not?

Exercise #22

Name_____

Section_____

Interpretation of Seismic Hazard and Geologic Quadrangle Maps

Objective: To understand how to read a published hazard assessment map and geologic map, and to use these maps to interpret potential hazards in the area surrounding Cal Poly Pomona.

Materials: ruler
calculator
California Geological Survey hazard assessment maps and reports (available in lab)
pencil
working knowledge of topographic map principles, geologic maps and profiles, groundwater conditions;
(Exercises previously completed from this manual)

Introduction

In this exercise, we will be using the State of California Seismic Hazard Zones map and the Geologic Map of the San Dimas 7.5' quadrangle topographic map. *Hazard maps* give information about the underlying geologic conditions and potential impact to existing and proposed development of the area. A 7.5' quadrangle (or quad) indicates that the map covers an area 7.5 minutes of latitude by 7.5 minutes of longitude. If your understanding of latitude, longitude, and 7.5' quadrangle maps is hazy, be sure to review **Exercise 9** in this laboratory manual.

The first group of questions (**Part A**) is based on the 1997 San Dimas, California, 7.5' Geologic quadrangle map. **Part B** utilizes the 1999 version of the San Dimas, California, 7.5' Seismic Hazard Zone quadrangle map and corresponding report.

Problems

Part A: Geologic Map of the San Dimas 7.5 Minute Quadrangle
1. Identify and describe the geologic units present on the Cal Poly Pomona campus. Include both the lithology and the corresponding age of the units.

2. What main geologic structural feature crosses the campus from east to west?

3. What is the main orientation of bedrock in the slope area just west of the old Administration Building (Building #1) at Cal Poly? This is the "L" shaped building with the flag. Note that this area has several specific orientations, so use your protractor to directly measure the structure symbols and express your answer as a range of strikes and dips.

4. What are the main faults located north of the San Jose Hills and how far away are they from campus (use the old Administration Building as a reference point)? Do these faults exhibit primarily strike-slip or dip-slip motion (based on the geomorphology)?

Part B: Seismic Hazard Zones

5. Identify which areas of the Cal Poly campus are potentially susceptible to *Liquefaction* hazard. What key geologic conditions of subsurface soil type and moisture are present to warrant such a hazard designation?

6. Identify which areas of the Cal Poly campus are potentially susceptible to *Earthquake-Induced Landslide* hazard. What key geologic conditions of subsurface soil type and moisture are present to warrant such a hazard designation?

7. Name at least three new campus structures and describe their subsurface hazards that are identified on the 1999 Seismic Hazards Zones map.

8. What is the approximate elevation of *Historically High Groundwater* conditions beneath the intersection of the I-10 and I-57 Freeways? Are there any well locations near this area to provide updated (current) results?

9. Look for the table of Shear Strength parameters in the Seismic Hazard Zone report. The geologic units are grouped on the basis of average *Angle of Internal Friction* (average Φ) and lithologic character. What "Groups" are represented by the geologic units on campus?

10. Give the range of shear strength parameters for the units on campus. Specify values for both Cohesion and Internal Friction. Fill in the table below; add rows if necessary:

Rock or Soil Unit	Cohesion	Angle of Internal Friction

Name _____

Class Number_____

Building in Landslide Country, Part 1

Objectives   Utilize geologic cross sections and geologic maps of sites with sloping ground to assess the potential for landslides and other types of mass movement or foundation failure.

Materials   pencil
colored pencils: green, yellow, red, blue, brown
lecture notes

Introduction

This exercise illustrates some negative aspects of building in areas where the ground has steep slopes or significant topographic irregularities. Students will learn to recognize unfavorable geologic conditions that might lead to devastating mass movements. In general, this will require knowing something about the relative strengths, porosity, and permeability of different Earth materials. The rest is just application of basic geologic intuition. Relevant reading material can be found in sections of your text that discuss mass movements and river erosion. Your lecture notes should help as well. Pay close attention to descriptions of the Vaiont, Italy dam disaster, the Point Fermin / Portuguese Bend, California landslides, various case histories of quick-clay failures and liquefaction (Turnagain Heights, Alaska and Rissa, Norway), and examples of cut slopes that expose day lighting planes of weakness.

As you work through the following problems, try to visualize each of the situations shown in three-dimensions, consider carefully the types of bedrock, sediment, or soil present, then apply a little common sense to identify potentially "bad situations". Use your colored pencils to highlight important areas. You may be surprised at the difference they make. Finally, consider the possibility that some of these examples could relate to your home!

Procedure / Problems

1. People choose to live in valleys for a variety of reasons, practical and aesthetic. However, most valleys exhibit relatively steep sides, and many form as the result of active river erosion. These conditions sometimes lead to problems for those people who choose to live on the relatively flat ground of valley floors. Other problems arise when the sloping sides of a valley are cut to create additional flat ground. Study the profile views of four different valleys (**Figures 20.1 a-d**). Then address the following problems by **writing directly on the relevant figure**:

A. Figure 23.1A shows tilted sedimentary rocks (sandstone, shale, and limestone). Assuming that a roadcut will be made into the slope, which side of the valley poses a greater slope stability threat? Explain.

B. Figure 23.1B illustrates two sets of planar fractures within relatively strong granite. Assess the stability of both sides of the valley. Explain your reasoning.

C. Figure 23.1C shows a river channel eroded into horizontal sedimentary rock layers. The cross section is located on a sharp bend in the river. The outside of this bend occurs on the east (right) side of the valley. Which rock layer is the weakest? What type of slope failure is imminent? Given that you must build on the sandstone, indicate the most secure place to locate your house.

D. Figure 23.1D shows faulted sedimentary rocks overlain in one area by an unconsolidated landslide deposit. Assess the stability of both sides of the valley. Explain your reasoning. Then select the best place to build. What kind of problems might you experience, even here?

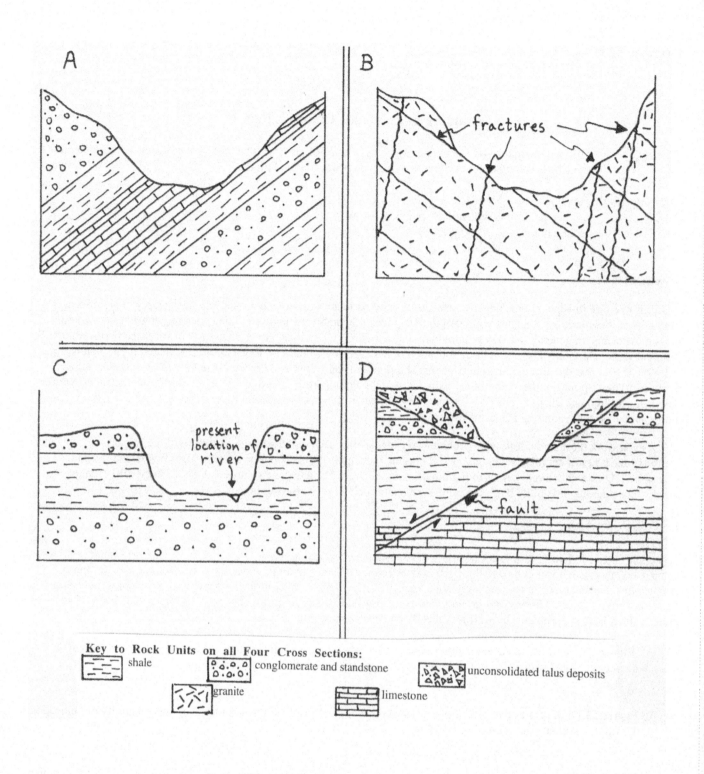

Figure 23.1. Transverse geologic cross sections through different valleys, illustrating potentially adverse slope conditions or foundation stability problems.

126

2. This problem addresses a hypothetical region that exhibits abundant slope stability issues as well as other geologic hazards. You may recognize several "bad situations" similar to examples given in lecture or in your reading. Study the geologic map (**Figure 23.2**) and accompanying cross sections (**Figure 23.3**). Refer to the legend and carefully consider the properties of the various rock types. Also study the map symbols and cross sections and try to visualize the orientation and geometry of potentially weak planes. Review of geologic mapping concepts in **Exercise #12** may be helpful. (Your instructor will provide additional lecture material to help you visualize geometric conditions leading to *Plane Failure* and *Wedge Failure*). Then address the following problems by writing directly on **Figures 23.2** and **23.3**.

A. Zone the map with respect to slope stability or foundation stability, using the following guidelines:

--Indicate stable segments of the highway with green, then use red to highlight segments subject to simple plane failure. Finally, use yellow to show roadcut segments subject to wedge failure.

--Along the river downstream from the dam, show stable and unstable portions of the limestone, assuming you plan to build near the limestone-shale contact along the river bank

--Clearly indicate all potential hazards on both sides of the reservoir. On which side of the reservoir (east or west) would you choose to position your waterfront dream home? Why? *Hint*: It may be helpful to research the Internet or your textbook with regard to historical disasters at Vaiont Dam, Italy or Saint Francis Dam, California.

B. Describe two likely scenarios that might result in dam failure and catastrophic flooding downstream from the reservoir.

C. List four potential geologic hazards that are intersected by cross section B-B'.

D. Locate a good site for your dream home somewhere on cross section B-B'. This should be a place that is reasonably flat and relatively free from geologic hazards.

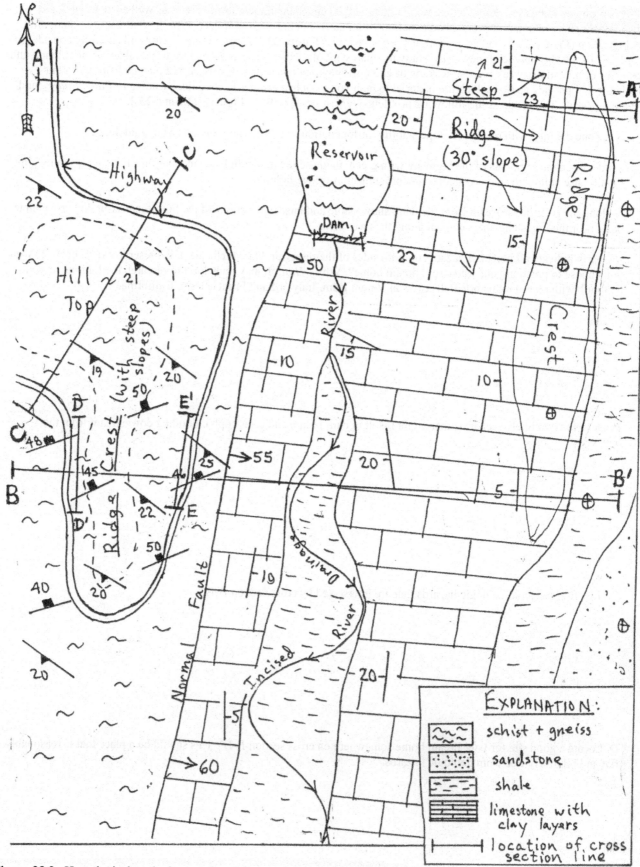

Figure 23.2. Hypothetical geologic map showing potentially unstable site conditions. Drawing by J. Nourse.

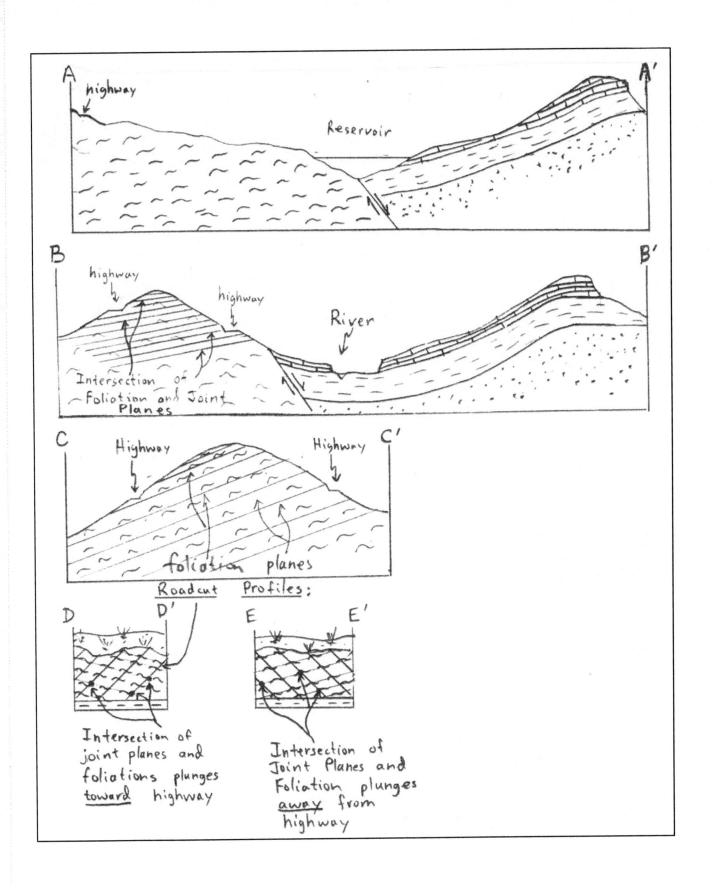

Figure 23.3. Geologic cross sections corresponding to the geologic map of **Figure 23.2.** Drawings by J. Nourse.

129

Name: _____

Class Number: _____

Building in Landslide Country: Part II

Objectives   To characterize recent landslide deposits and measure the extent of damage; identify and delineate pre-existing landslides using 3-dimensional imagery software; and use that imagery to assess the given level of risk and establish a building setback zone.

Materials   Google Earth or other Computer 3-D imagery software;
Word processor software; lecture notes

Introduction

In this exercise we will continue working with landslides and build upon your work from **Exercise 23**. We will use 3-dimensional imagery software (Google Earth) to characterize recent landslides in La Conchita, CA and assess the area for further landslide potential. The first part of this exercise will review 1995 and 2005 landslide deposits in southern California and measure their extent & impact.

Landslide classification is based on a wide variety of motion, including sliding, spreading, flowing, falling or toppling. **Figure 24.1** shows a generalized cross section view of an earth flow failure. **Table 24.1** describes a generalized classification system for identifying the various types of landslides.

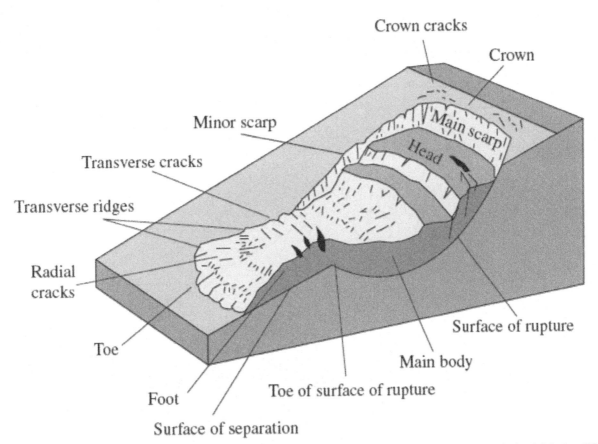

Figure 24.1: An idealized slump-earth flow showing commonly used nomenclature for labelling the parts of a landslide (modified from USGS FS 2004-3072).

TYPE OF MOVEMENT		TYPE OF MATERIAL		
		BEDROCK	ENGINEERING SOILS	
			Predominantly coarse	Predominantly fine
FALLS		Rock fall	Debris fall	Earth fall
TOPPLES		Rock topple	Debris topple	Earth topple
SLIDES	ROTATIONAL	Rock slide	Debris slide	Earth slide
	TRANSLATIONAL			
LATERAL SPREADS		Rock spread	Debris spread	Earth spread
FLOWS		Rock flow (deep creep)	Debris flow	Earth flow (soil creep)
COMPLEX		Combination of two or more principal types of movement		

Table 24.1. Types of landslides. Abbreviated version of Varnes' classification of slope movements (Varnes et. al., 1978)

Procedure / Problems

Start Google Earth, then download and open the **Exercise 24 kmz file** on the CPP blackboard system.

Problem #1 – Initial Observations & Site Characterization:

The 1995 landslide in La Conchita was classified as an "Earth Flow" or "Debris Flow"—see also **Figure 24.2** below. Be sure to review your Landslide Handbook and understand why it is classified as such. Refer to the following link:

https://pubs.usgs.gov/circ/1325/

Using the Google Earth toolbar, measure and document the maximum length & width of damage at the top of the 1995/2005 landslide affected area. Measure both in feet & in meters. Also measure and document the maximum width of damage at toe of the slope. Note that there is one scar area but two separate toes. This observation will become important later on in the exercise when we assess the overall hazard potential.

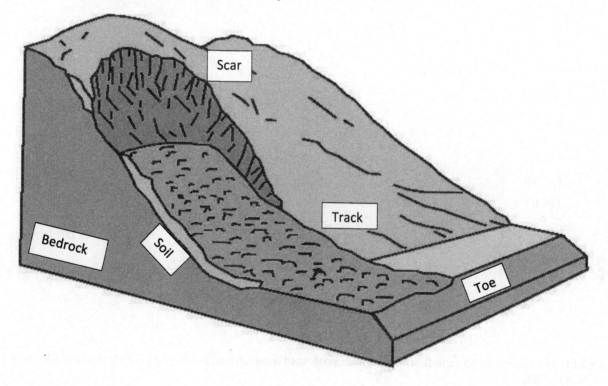

Figure 24.2. An idealized Earth flow schematic (as modified by Highland and Bobrowsky, 2008).

132

Problem #2 – Characterizing the Landslide areas:

Using the "Polygon" feature on the Google Earth toolbar, highlight, identify and label the key parts of the 1995 & 2005 La Conchita earth flows (you can use MS Word or other similar software). Copy & Paste the figure as part of your answer. Using your Word processor software, add clear labels and arrows identifying the various features. Remember that flows generally have wide scar and toe regions, often on the order of tens to hundreds of meters across, but narrow tracks where they travel down to lower elevations, often on the order of just a few meters.

Accessing Historic Imagery:

Historical imagery can be used to identify changes that have occurred in a site over time. These changes can then be used in your assessment of potential hazard areas. Google Earth is an effective tool that can help with this task. Scale back and rotate your Google Earth view so that you are positioned vertically above the site and access the 1994 imagery. Do this by selecting the clock icon in the toolbar. A new window should show up at the top left of the image. Slide the bar to 1994. The background image should change to reflect the corresponding image date (look below the image to see the exact date). As you switch back and forth through the various historical images, you should now be able to see the portions of the La Conchita community buried by the 1995 slide.

Problem #3 – Potential Setback Zones:

Estimate the distance from the original (pre 1994 slide) toe of slope to the post-landslide toe of slope. Do this for both landslides. These measurements estimate how far the landslide penetrated into the community. You can use the "Placemarks" feature in the toolbar and/or comparison with manmade structures nearby and unaffected by the landslides to help with your measurements. Use this distance to establish a "*Minimum Potential Setback Zone*" for future landslide hazards.

Effects of previous landslides: Previous landslides can strongly influence vegetation on a hillside. This can be a valuable clue for identifying pre-existing landslide areas that may move in the future. You can see this phenomenon by reviewing historic imagery and identifying vegetation conditions. Look at the 1995 and 2005 landslide areas and compare them to the adjacent slope areas on either side. Notice the vegetation color in the historical imagery. You should be able to see areas of both brown and green vegetation. For example, look between August 2006 (dry conditions) & April 2011 (wet conditions). Wet season months are characteristically green, meaning the overall hillside vegetation is healthy. Yet during the dry season, most of the hillside vegetation turns brown as the annual grasses die off. The dry-season images consistently show several patches of year-round green vegetation. Think about why this is happening.

Problem #4 – Landslide Moisture Characteristics:

Find the best and sharpest wet-season and dry-season imagery you have available for review. You should see areas along the surrounding hillside that annually transition from green to brown. You should also see that the track & toes of the 1995 & 2005 landslides stay green longer than the rest of the slope, with some areas remaining green year-round. Identify these areas and explain why would the earth flow fan (mound of deposition) be greener than the surrounding areas? (Hint: think of the overall in-place bedrock in terms of porosity and permeability; then think of the landslide debris porosity and permeability; how would these parameters differ for each geologic unit?).

Problem #5 – Identifying Other Potential Landslides:

Use the mouse or Google Earth controls to zoom in/out and identify to other potential landslide areas on the slope behind the La Conchita community. Create a candidate list of suspected landslides based on green dry-season vegetation where green mounds are located within brown patches. To help refine your list look for small characteristic scars above the dry-season green areas. This is strong evidence of other landslides, in addition to the 1995/2005 ones. How many total did you find? Copy/Paste an image of the overall hillside and circle the additional suspected landslides as part of your analysis.

Problem #6 – Delineate Additional Landslides on the Slope:

Use the mouse or Google Earth controls to zoom in/out and identify to other potential landslide areas on the slope behind the La Conchita community. Create a candidate list of suspected landslides based on green dry-season vegetation where green mounds are located within brown patches. To help refine your list look for small characteristic scars above the dry-season green areas. This is strong evidence of other landslides, in addition to the 1995/2005 ones. How many total did you find? Copy/Paste an image of the overall hillside and circle the additional suspected landslides as part of your analysis.

Advanced Problem – Understanding the Ancient landslide Complex:

Google the following document: USGS Open File Report 2005-1067. Look at the description for the "ancient landslide" that involved the entire bluff. You should find an oblique false-color infrared image taken in 2002 and an oblique LIDAR image or the bluff taken just after the 2005 landslide. Find these features in your Google Earth imagery. There is an even larger suspected (but as yet unproven) landslide behind the "ancient landslide".

Name: _____

Class Number: _____

Grading Plans; Cut and Fill Procedures

Objectives   This exercise covers a portion of site development known as grading plans. Students will learn to:
(a) Understand and read grading plans; (b) Draw quick and accurate profiles showing before and after conditions; (c) Identify cut/fill lines; and (d) Make rough-estimate earth material volume calculations.

Materials   Pencil and calculator; engineering graph paper; black pen; red and green colored pencils
Word and Excel software; lecture notes
Access to supplementary materials on Blackboard

Introduction

Grading plans are diagrams where existing topographic conditions and proposed development are superimposed onto the same illustration. A portion of a grading plan is shown in **Figure 25.1** below. These diagrams are utilized through all phases of development and guide the preliminary investigation and the overall project. Once the preliminary investigation is concluded, the grading plans are typically revised to include initial site development observations and subsequent geotechnical recommendations. Grading plans often go through a series of revisions where the design process is refined and the final product is achieved. After the construction phase of the process is concluded, a final "*As-Graded*" or "*As-Built*" plan is issued to record the finalized site topography.

Figure 25.1. Example of a grading plan showing a proposed highway crossing an existing canyon. Contour values are in feet above sea level. Distance between + marks is 200 ft. Read below for other details. Drawing provided by E. Roumelis.

Procedure / Problems

This exercise combines several skills you have developed from other labs finished during this past quarter. Take a look at the "AS-BUILT" grading plan example in the lab classroom. Note the specific changes that were hand-written and saved as the final exhibit recorded in the County and/or City offices of Building and Safety. Many of the previous exercises this quarter has been intended for you to develop and hone the necessary skills handle this lab assignment. Pertinent labs are noted in each problem where you developed the skills needed to complete the tasks below.

The grading plan in **Figure 25.1** above shows existing topography as fine, thin dotted contour lines (similar to what you worked on in **Exercises #9 and #10**). The grading plan also shows a second layer of topographic lines, with a thicker bold line and a slightly different font for a second set of elevation contour line values. The thicker bold lines show "proposed elevation" or what the final design will eventually become. Utilizing multiple contours (surfaces) takes some getting used to. Dealing with multiple contour images on the same map was a similar concept you encountered in the groundwater **Exercise #17**. Your instructor will review these concepts, using **Figure 25.1** to demonstrate procedures that you will apply to a larger grading plan.

1. Cut and Fill: The overall grading plan (see *Exercise 25 Grading Plan.PDF* posted on **BlackBoard**) depicts a proposed roadway design through a series of granitic hills. Please open this PDF file. Identify areas of "<u>**CUT**</u>" (where material will be removed to achieve the proposed elevation) and "<u>**FILL**</u>" (where material will be added to achieve the proposed elevation). To do this, *simply pick any point within the bold area and compare existing versus proposed elevation*. **Lightly shade** the cut areas in **red** and the fill areas in **green**. Also, **highlight with a black line** the "cut/fill" line (boundary between these areas) clearly on your map.

2. Draw the profiles shown on the map. You have made several profiles so far in **Exercises #10, #12, #13, and #14**. Use the gridded side on your green engineering paper, and remember to use the upper margins to transfer the data quickly from the map to the profiles. Each profile should take you 5 minutes **MAXIMUM.** Remember, in the industry time is money; so time yourself to see how far you have progressed. If you find that this task is taking you too long please arrange to meet with me so that I can help speed up your process. Also, watch for the "benches" in the proposed slope configurations. *NOTE:* the "+" marks on the map are 200 feet apart from each other in the X-Y direction). Do not use vertical exaggeration for the cross section as this will give you a false perception. Estimate the steepness of the proposed slopes using standard industry terminology (1:1 slope, 2:1 slope, etc.).

Quantity calculations are used to help your client estimate their grading and development costs for a particular project. This task is performed by the Civil or Geotechnical Engineering team at several stages of a project lifespan. The next two problems use quantity estimates (see **Exercises #11 and #13** for different approaches).

3. Estimate the total volume of material to be removed in each cut area and compare this to the total volume of material to be filled. Name each cut and fill region (Cut A, Cut B, Fill C) & tabulate the individual volumes. Express your answers in **cubic yards** (the industry standard). **Create a summary table of your results.**

4. In grading operations, difference in quantity of earth material between the amounts of cut and fill is calculated. This is done on an individual area basis within large grading jobs as well as a total for the entire site. Grading contractors then use this information to plan their schedule more efficiently. The goal is to move the greatest amount of material the shortest distance in the least amount of time, thus minimizing grading costs. When quantifying earthwork values, the comparative descriptions are called "*Long*", "*Short*" or "*Balanced*". "*Long*" describes a situation where more cut exists than fill. When a cut/fill comparison is deemed "*Long*" left-over earth material must be exported (at a significant cost to the client) to other areas or grading projects to complete the grading project. "*Short*" describes a condition where there is more fill than available cut. When a cut/fill comparison is deemed "*Short*" the fill quantity is greater than the cut and importing of additional soil from other grading projects areas is required (again at a significant cost to the client). "*Balanced*" is the condition where cut and fill are ~equal (usually within 5% to 10% of the quantity total). Many engineering projects have ended up in costly litigation over cut-fill quantity errors and corresponding project cost overruns. In **Problem 3** above, you tabulated overall earthwork cut and fill operations on the same road alignment. **Was there a raw balance? If not, were the cut/fill values long or short? By how much?**

136

Shrinkage & Bulking: During a grading operation, excavation of earth material is made (cut), the material is then transported by earth working equipment (scrapers, dump trucks, etc.) and deposited in a fill area and subsequently compacted. This creates 3 different conditions called _Bank_, _Loose_, and _Compacted_ as shown below in **Figure 25.2**.

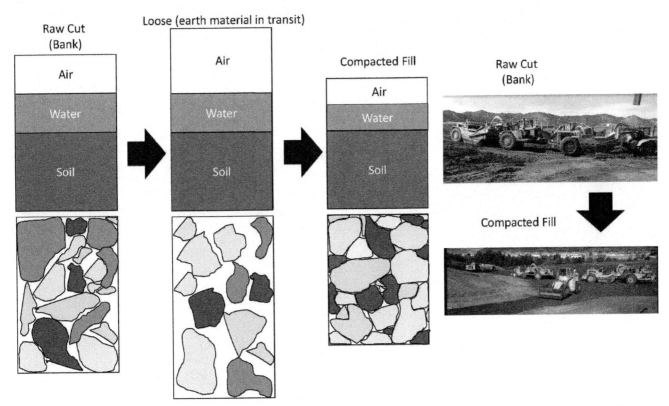

Figure 25.2. Visual depiction of earth material during a grading operation. Note that the example above shows an alluvium-cut region. Denser cuts (such as in crystalline bedrock) could easily yield an overall bulking effect. Drawings and photographs by E. Roumelis.

It is the design team's responsibility to assess values of shrinkage & bulking as the earth material is transported from the initial (**cut**) to the final (**fill**) location. The initial state or "_Bank_" condition is evaluated during the investigation phase of a project. This value is determined by geotechnical testing of "in-situ" sampling in proposed cut areas. During the excavation process also referred to as the "_Loose_" condition, the earth material is excavated, ripped up, disturbed, in a loose granular state, and therefore has significantly **lower density**. The lower density means the transported earth material occupies a greater volume. Upon destination to the fill area, the material is dumped and spread out in thin, loose lifts. Since loose earth material is susceptible to consolidation, the material must be compacted prior to construction of settlement-sensitive structures. The "_Compacted_" condition is often a targeted minimum level of compaction. The level of compaction varies based on what type of structure is proposed; but typically 90% of _Modified Proctor density_ is a minimum compaction requirement.

Watch the following short video references on Standard versus Modified Proctor tests and field compaction testing methods: **https://www.youtube.com/watch?v=6ZvmvPwlDSc**

5. Shrinkage & Bulking: In this exercise we are dealing with a granite site. Granite is typically very dense and with near zero air voids in its natural state. A complete S&B analysis is beyond the scope of this lab, so we will simply summarize by estimating the final condition between "_Bank_" & "_Compacted_" as a **bulking of 10%**. Recalculate your quantities in **Problem 5** by allowing a Bulking factor of 10% between the cut & fill. Does this change your overall balance calculation? How would you correct this? (_Hint: consider the roadway elevation and how the amounts of cut & fill change when you raise or lower the roadway_).

6. According to your tabulations, describe what would you do to achieve quantities within 5% of balance? Be specific in your recommendations. (_Hint: What happens to your overall cut/fill quantities if you move the roadway up, down or laterally across the page?_)

Bibliography

Abbott, Patrick L., 2002, *Natural Disasters, 3rd Edition,* McGraw-Hill Companies, 422 p.

Anderson, R.S., 1990, Evolution of the northern Santa Cruz Mountains by advection of crust past a San Andreas Fault bend: Science, v. 249, p. 397-401.

Anderson, R.S., and Menking, K.M., 1994, The Quaternary marine terraces of Santa Cruz, California: Evidence for coseismic uplift on two faults: Geological Society of America Bulletin, v. 106, p. 649-664.

Blatt, Harvey and Tracy, Robert J., 1996, *Petrology: Igneous, Sedimentary and Metamorphic, 2nd Edition*, W. H Freeman and Company, New York, 529 p.

Bolt, Bruce A., 1978, *Earthquakes: A Primer,* W. H. Freeman and Company, New York, 158 p.

Boore, D. M. and Atkinson, G. M., 2008, Ground-motion prediction equations for the average horizontal component of PGA, PGV, and 5%-damped PSA at spectral periods between 0.01s and 10.0s, Earthquake Spectra, *24(1): 99–138. doi: 10.1193/1.2830434.*

Borcherdt, R. D. (ed.), 1975, Studies for seismic zonation of the San Francisco Bay region. *U. S. Geological Survey Professional Paper 941-A*, 102 p.

Bradley, W.C., and Griggs, G.B., 1976, Form, genesis, and deformation of central California wave-cut platforms: Geological Society of America Bulletin, v. 87, p. 433-449.

Coduto, D. P., Young, M. R., and Kitch, W. A., 2011, *Geotechnical Engineering: Principles and Practices*, Pearson Education, 760 p.

Fetter, C. W., 2001, *Applied Hydrogeology, 4th Edition*, Prentice-Hall, Inc., 598 p.

Greensfelder, R. W., 1980, *California Division of Mines and Geology Special Report 120*, plate 1B.

Hamblin, W. K., and Howard, J. D., 1999, *Exercises in Physical Geology, 10th Edition*, Prentice-Hall, Inc., 256 p.

Highland, L., *Landslide types and Processes, U. S. Geological Survey Fact Sheet 2004-3072, July 2004,* 4p.

Highland, L., and Bobrowsky, P., *The Landslide Handbook – A Guide to Understanding Landslides, U. S. Geological Survey Circular 1325, November 2008,* 147p.

Jibson, Randal, W., *Landslide Hazards at La Conchita, California, U. S. Geological Survey Open File Report 2005-1067,* 12p.

Keller, Edward .A., 1992, *Environmental Geology* (6th Edition), Macmillan Publishing Co., 517 p.

Lajoie, K.R., 1986, Coastal tectonics, in Studies in Geophysics, Tectonics: National Academy Press, Washington D.C., p. 95-124.

Meritts, Dorothy J., De Wet, Andrew, and Menking, Kirsten, 1998, *Environmental Geology: An Earth Systems Approach*, W. H. Freeman and Company, 452 p.

Morton, Douglas M., Sadler, Peter M., and Minnich, Richard A., 1987, Large rock avalanche deposits: examples from the central and eastern San Gabriel Mountains of southern California, in *Field guide to landslides of the inland valleys and adjacent mountains of southern California,* Publications of the Inland Geological Society, Volume 32, Part III, p. 20-32.

Myles, Douglas, 1985, *The Great Waves,* McGraw Hill, 202 p.

Open Topography, 2012, Wallace Creek to Phelan Creek along the San Andreas Fault, Carrizo Plain, California (50 cm DEM), available at: https://drive.google.com/file/d/0B2YW8UjIIMEhNkNWOHVjMl9kUFU/view

Perg, L.A., Anderson, R.S., and Finkel, R.C., 2001, Use of a new 10Be and 26Al inventory method to date marine terraces, Santa Cruz, California, USA: Geology, v. 29, no. 10, p. 879-882.

Poort, Jon M., 1980, *Historical Geology: Interpretations and Applications* (3rd Edition), Macmillan Publishing Co., 182 p.

Rossbacher, Lisa A., Jessey, D. R., and Berry, D. R., 1986, *Earth Science Exercises*, Kendall/Hunt Publishing Co., 80 p.

Sieh, K., and Jahn, R.H., 1984, Holocene activity of the San Andreas fault at Wallace Creek, California: Geological Society of America Bulletin, v. 95, p. 883-896.

Sieh, K., and Wallace, R.E., 1987, The San Andreas fault at Wallace Creek, San Luis Obispo County, California: Geological Society of America Centennial Field Guide – Cordilleran Section, p. 233-238.

Southern California Earthquake Center (SCEC), 2008, Wallace Creek Interpretive Trail: A geologic guide to the San Andreas Fault at Wallace Creek, available at: http://scecinfo.usc.edu/wallacecreek/index.html

Stover, C.W., and Coffman, J.L., 1993, Seismicity of the United States, 1568-1989 (Revised): U.S. Geological Survey Professional Paper 1527, 418 p.

Toro , G. R. , Abrahamson , N. A. and Schneider , J. F. , 1997, Model of strong ground motions from earthquakes in the central and eastern North America: Best estimates and uncertainties, Seismological Research Letters, v. 68, no.1, p. 41 – 57.

Varnes, D.J., 1978, Slope Movement types and processes, *in* Schuster, R.L., and Krizek, R.J., eds., Landslides – Analysis and Control: National research council, Washington D.C., transportation Research Board, Special Report 176, p 11-33.

Vedder, J.G., and Wallace, R.E., 1970, Map showing recently active breaks along the San Andreas and related faults between Cholame Valley and Tejon Pass, California: U.S. Geological Survey Miscellaneous Geologic Investigations I-574, 2 sheets, scale 1:24:000.

Weber, G.E., 1990, Late Pleistocene slip rates on the San Gregorio Fault Zone at Point Año Nuevo, San Mateo County, California, in Garrison, R.E., Greene, H.G., Hicks, K.R., Weber, G.E., and Wright, T.L., eds., Geology and Tectonics of the Central California Coast Region, San Francisco to Monterey, Volume and Guidebook, Pacific Section, American Association of Petroleum Geologists, Bakersfield, California, p. 193-204.

Weber, G.E., and Allwardt, A.O., 2001, Field Trip 1: The Geology from Santa Cruz to Point Año Nuevo—The San Gregorio Fault Zone and Pleistocene Marine Terraces. in. P. W. Stoffer and L. C. Gordon, eds., pp. 1-32, Geology and Natural History of the San Francisco Bay Area: A Field-Trip Guidebook 2001 Fall Field Conference National Association of Geoscience Teachers Far Western Section September 14–16, 2001 Menlo Park, California. Bulletin no. 2188. U.S. Geological Survey, Reston, Virginia.

Winchester, Simon, 2001, *The Map that Changed the World: William Smith and the Birth of Modern Geology*, Harper Perennial, 352 p.

CPSIA information can be obtained
at www.ICGtesting.com
Printed in the USA
LVOW02s2332260517

535099LV00010B/15/P